M

Distributions of Correlation Coefficients

Hoben Thomas

Distributions of Correlation Coefficients

With 17 Illustrations

Springer-Verlag
New York Berlin Heidelberg
London Paris Tokyo

Hoben Thomas
Department of Psychology
Pennsylvania State University
University Park, PA 16802
USA

Library of Congress Cataloging-in-Publication Data
Thomas, Hoben.
  Distributions of correlation coefficients / Hoben Thomas.
  p.   cm.
  Bibliography: p.
  Includes indexes.
  1. Occupational aptitude tests—Evaluation—Statistical methods.
2. Examinations—Validity—Statistical methods.   3. Correlation
(Statistics)   I. Title.
HF5381.7.T48   1989
153.9′4—dc19   88-38968

Printed on acid-free paper.

Typeset by Asco Trade Typesetting Ltd., Hong Kong.
Printed and bound by Edwards Brothers Incorporated, Ann Arbor, Michigan.
Printed in the United States of America.

9 8 7 6 5 4 3 2 1

ISBN 0-387-96863-6 Springer-Verlag New York Berlin Heidelberg
ISBN 3-540-96863-6 Springer-Verlag Berlin Heidelberg New York

To Doris and Patricia

Preface

This monograph is motivated by a problem in personnel psychology, more specifically, a problem in psychometrics called validity generalization. From a statistical perspective, the essence of the problem is to make statements about population correlation coefficients, ρ_i, inferred from a collection of sample correlation coefficients, R_i, $i = 1, \ldots, k$. Procedures of validity generalization have largely been concerned with estimating the variability of ρ_i. The first part of the book examines the largely ad hoc procedures that have been used rather widely over the last decade. The second part develops a new model formulated from the perspective of finite mixture theory and illustrates its use in several applications.

The monograph has been substantially improved because of the many useful comments contributed by Steven F. Arnold, Department of Statistics, Pennsylvania State University; Philip Bobko, Department of Management, University of Kentucky; Fritz Draskow, Department of Psychology, University of Illinois; Frank Landy, Department of Psychology, Pennsylvania State University; and Rolf Steyer, Department of Psychology, University of Trier, FRG. I appreciate their efforts. Any errors or mistakes are, of course, my own.

University Park, PA   Hoben Thomas

Contents

1
Introduction

1.1 Motivation and Background

The goal of this book is to understand histograms, such as Figure 1.1. The figure is taken from Ghiselli's classic 1966 book *The Validity of Occupational Aptitude Tests* and is his Figure 2-4. It shows histograms of observed correlation coefficients called validity coefficients. The problem is to model such histograms. Specifically, what might be a parent distribution for such histograms? How many different population correlation coefficients, if more than one, might have given rise to the data? How variable might be the population correlation coefficients? Can their values be estimated?

A fundamental activity is personnel psychology is to develop predictive measures of job performance from tests. Such activities are often termed criterion validation studies, or briefly, validity studies. Depending on the setting, the test may be specifically developed for a job, or it might be a standard, readily available instrument. The performance measure might be a job training measure, a supervisor's proficiency rating, or some simple measure of work performance. Usually, a simple model relating a performance measure and a predictor variable is considered adequate, and the performance measure is linearily regressed on the predictor variable. If the regression function obtained is found to usefully relate test and criterion performance, the test is often subsequently used as a selection device; future job applicants with suitable scores, typically those persons with scores above some predictor variable critical value, are selected.

Although it might seem conceptually more natural to consider the adequacy of test prediction from the viewpoint of a regression model, personnel psychologists have historically viewed the problem from a correlational perspective and have taken as the predictive summary indices correlation coefficients (validity coefficients). The choice of the correlation coefficient is not necessarily a bad one since its size alone does indicate how well the variables are linearly related.

According to Burke (1984), thousands of validation studies are performed yearly in the United States. If the cost of a validation study for a medium-size

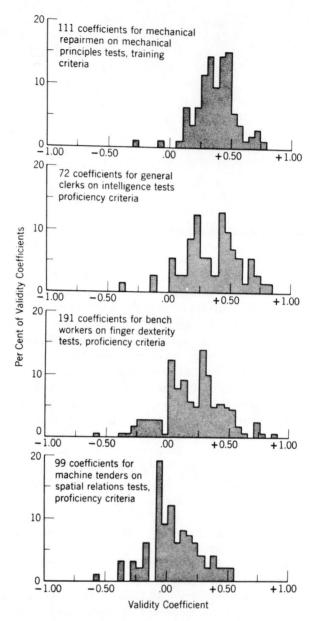

FIGURE 1.1. Histograms of validity coefficients for four occupational groups. (From *The Validity of Occupational Aptitude Tests* (p. 29) by E.E. Ghiselli, 1966, New York: John Wiley & Sons, Inc. Copyright 1966 by John Wiley & Sons, Inc. Reprinted by permission.)

company is about $30,000 (Landy, 1985, p. 68), clearly a substantial amount of time and money is involved. But, the results of these efforts have rarely been flattering. Figure 1.1 shows why. The validity coefficients are usually not very high. Depending on the setting, the correlation coefficients may average .45 or so, and sometimes are strikingly less. More troublesome is the large variability among the validity coefficients for seemingly similar jobs with similar performance criteria. Such results led to the perspective that a test's predictive validity was highly specific to each job setting, necessitating that validation studies be performed afresh in each new case. This situational specific viewpoint, which was Ghiselli's (1966) viewpoint, dominated thinking for decades.

Things changed however during the mid 1970s with the publication by Schmidt and Hunter (1977). They argued that tests and their associated prediction equations were much more generalizable from setting to setting than had been believed. Thus, it was claimed validity generalization was possible. There was no need to do a fresh validition study each time a new job selection problem was confronted. Results obtained from a related setting might serve quite well (Note 1.1).

Schmidt and Hunter proposed no new method of testing; rather they argued that the size of validity coefficients was small and their variability large because of statistical variables—typically termed artifacts—such as variation due to sample size, test and criterion measurement unreliability, and so forth. Such variables have long been known to influence validity coefficients, and through classical test theory conceptualizations (e.g., Lord & Novick, 1968), it is possible to correct individual correlations for these artifacts. What Schmidt and Hunter purported to do was to provide a comprehensive, often computationally intensive, procedure to correct, en masse, samples of correlation coefficients, such as in Figure 1.1—purging them, if you will, of these artifacts—so that it could be determined how much variation in the population correlation coefficients remained after these corrections were applied. Once the corrections have been made, it has been argued, there often remains little variation in the correlation coefficients. Presumably therefore, the large variability among the correlation coefficients observed in histograms, such as in Figure 1.1, is due not primarily to factors unique to individual testing situations but rather to the artifacts that are always by-products of any assessment situation. If, therefore, the observed variability of sample correlations is attributable to artifacts and not true variation among testing settings (that is, variation among the population correlation coefficients associated with each testing setting), validity generalization is possible.

Since the Schmidt and Hunter paper (1977), there has been considerable research activity concerned with validity generalization. A large portion of the literature has been contributed by Schmidt, Hunter, and their colleagues. Virtually all of the validity generalization literature is found in two applied journals: primarily the *Journal of Applied Psychology* and secondarily *Personnel Psychology*, which recently devoted an entire 100-plus page issue to the topic (Schmidt, Hunter, Pearlman, & Hirsh, 1985; Sackett, Schmitt, Tenopyr, Kehoe, & Zedeck, 1985). This issue also provides an extensive bibliography.

Collectively, this literature is largely empirical, often with a heavy computational focus, either applying such methods to samples of correlation coefficients or using simulation in an effort to understand the procedures.

Certainly, from within the field of personnel psychology and applied testing, Schmidt and Hunter's work is widely known and respected. For instance, Anastasi calls their work statistically "sophisticated" and maintains that "the validity of tests of verbal, numerical, and abstract reasoning aptitudes can be generalized far more widely across occupations than has heretofore been recognized" (1986, p. 11). Others have suggested "that test validities are broadly generalizable across applicant populations, geographic locations, and jobs and even 'gross changes' or differences in job tasks do not destroy validity" (Baker & Terpstra, 1982, p. 604). Some persons apparently believe that things are so well in hand, testing programs can be routinized by clerical staff. "One can even go so far as to suggest that practitioners and researchers will only need to analyze jobs or situations of concern, and on the basis of these analyses consult tables of generalized abilities...." (Zedeck & Cascio, 1984, p. 486).

The concern here is *not* with whether validity generalization is possible or not. Even from a naive perspective it would certainly be expected that given two jobs with similar descriptions and worker populations, when using similar tests and criteria, similar regression functions should be obtained. Rather, a main goal is to probe analytically the validity generalization model and the statistical estimation methods. It seems critically important to do so. The application of validity generalization procedures is being extended in many practical domains where its influence on the employment prospects of job applicants can be critically important.

Furthermore, it has been reported that many large corporations or organizations, such as the Central Intelligence Agency, Proctor and Gamble, and Sears, have used validity generalization procedures, as have 13 large petroleum companies and the US Employment Service according to Schmidt et al. (1985). They also note (1985, p. 707) that it has been recommended that validity generalization procedures be part of the revised federal standard *Uniform Guidelines on Employee Selection Procedures* (1978), which is one of the key documents dictating testing procedures and thus is critical for litigation. In addition, federal district courts have tended to render decisions supporting validity generalization (Sharf, 1987).

1.2 Conceptual Problems of Validity Generalization

From an analytical perspective, there are several observations about the status and development of validity generalization procedures that seem especially noteworthy, particularily in light of their apparently pervasive applications in business, industry, and government.

1. *No carefully defined probability model for validity generalization has ever appeared* (Note 1.2). In their original 1977 paper, Schmidt and Hunter proposed no formal model; they sketched the basic ideas behind their proposal and suggested a procedure. This situation remains largely unchanged although efforts have been made toward providing such a model (Callender & Osburn, 1980).

2. After ten years of research activity, *there are no theorems or general results that pertain to the validity generalization model (to the extent there is a model) or to any of the several different but closely related estimation methods available.* Yet many fundamental questions have remained unanswered for years, questions that are addressable through analysis. For example, it remains unclear precisely *what* the procedures estimate.

A standard procedure for understanding a statistical estimator is to consider the expectation of the estimator. Yet, there appears to have been no effort to proceed in this way. Furthermore, no optimality criteria for the estimation procedures have been given. For example, what are the solution criteria? Is there a function to be maximized?

Clearly, one reason why uncertainty remains, particularily with respect to the issue of estimation, is because there has not appeared a satisfactory probability model amenable to analysis. Consequently, the estimation procedures have not flowed naturally from a formal model; rather, they have developed in an ad hoc manner. This fact has undoubtedy contributed to the proliferation of simulation studies that have replaced analysis.

Doubtlessly, a second reason why no general model properties have been provided is that it has been believed that there is nothing special about validity generalization. Hunter, Schmidt, and Jackson (1982, p. 43) regard certain correction procedures underlying validity generalization as exactly classical test theory. Presumably therefore, general results characterizing classical test theory apply with full force to validity generalization procedures. As will be seen, this appears not to be the case.

3. It seems extraordinary that given the foundational preeminence of empirical validity coefficient histograms, such as Figure 1.1, *there has never appeared a suggestion as to what population model might give rise to such correlational data*, at least no suggestion could be found in the literature. Precisely what parent distributions have given rise to the panels of Figure 1.1? That the question seems not to have been raised let alone answered seems remarkable.

Given that virtually all inferential statistical methods start with *some* assumptions regarding a population distribution from which the observations have arisen, one might reasonably raise the question: How well developed is validity generalization and what could estimates obtained from the procedures be estimating? Chapters 2 and 3 speak directly to these issues. The answer to be given in these chapters is that even when satisfactorily defined, the model has very peculiar properties that largely prohibit the development of rigorous estimation procedures. Regarding estimation, there may be times when the procedures lead to sensible conclusions. But, the estimators lack virtually all of the desirable properties of more familiar estimation procedures, and the estimators can display clearly pathological properties.

1.3 An Alternative Formulation

Because the model and methods of validity generalization will be judged to have fatal shortcomings, a second goal of this book is to try to put the generalizability versus specificity issue on a more rigorous footing.

The central problem of interpreting Figure 1.1 is to infer the number of values of the population correlation coefficients that give rise to such histograms. To do so, a new model is proposed. The basic idea is to view the problem from the perspective of finite mixture theory (e.g., Everitt & Hand, 1981). From this perspective, the panels of Figure 1.1 are viewed as samples from mixture distributions. The components of this mixture distribution are distributions of R, the correlation coefficient, that are assumed to arise from the bivariate normality of the predictor and criterion variables. The theory that produces this conclusion is developed in Chapter 5. Chapter 6 develops maximum likelihood estimation procedures for the population correlation coefficients and their associated weights when the number of components in the mixture is specified. Chapter 7 provides several numerical examples based on published data.

Under the model, it is possible to provide fitting procedures for evaluating the model given the histogram data. Using resampling procedures, the variance–covariance matrix of the model estimates can be given. Thus, approximate confidence intervals and tests for the parameters can be constructed. Examples of these and related procedures are also provided in Chapter 7.

The model explicitly includes sample size as a known model parameter, so this artifact is always considered in any analysis. The problem of correcting for unreliability and range restriction is considered, along with remaining issues, in Chapter 8.

The book is in two distinct parts: Chapters 1 through 4 consider the existing model and methods; Chapter 2, which parameterizes a validity generalization model and explores its properties states several general results. This chapter is the heart of the first part of the book. Chapter 4 is in part a self-contained general summary of the main findings regarding the validity generalization model (as parameterized here) and estimation methods. Chapters 5 to 8 develop the conditional mixture model and associated estimation methods.

2
A Validity Generalization Model

2.1 Introduction

The validity generalization literature is difficult to read. For persons comfortable with the development of a typical article in, for example, *JASA*, *Technometrics*, or *Psychometrika*, this is especially so. Such articles usually have a clear statement of a probability model. The model's properties are examined, then the estimation problem is considered as a conceptually distinct problem. Unfortunately, these conceptual and organizational features do not characterize the validity generalization literature. For example, one searches to find a clear presentation of the validity generalization model.

As remarked previously, the original Schmidt and Hunter (1977) paper presented no model. It was essentially a prescription for estimation of true validity variance, the meaning of which will be clear subsequently. A better source is Hunter et al. (1982, p. 42 and following), but the development is not satisfactory. For example, variances of additive compositions are considered with the apparent assumption that the random variables are independent or uncorrelated, but no statement is given as to which assumption is made or which, in fact is even plausible or possible (p. 43). Apparently, fictitious, examples are provided in which the estimated variance components are negative. No comment is provided about them (pp. 45, 72). Such negative estimates may be nothing more than picadillos to be sure; but one cannot be certain without more careful analysis. Whatever, such findings are not reasurring. No properties of the model nor of the estimators are considered.

Callender and Osburn (1980) can be credited with the first effort to specify a precise model, and in fact, some of the development of Hunter et al. (1982) follows their lead. However, Callender and Osburn's (1980) development is not clear. Fixed constants in one equation become random variables in the next (compare their Equations 7 and 8) with neither conceptual nor notational clarity, and the model and estimation problems become intertwined. No general properties of the model are stated, nor are any analytical properties of the estimators considered.

The conceptual heritage of validity generalization appears to come from two primary sources: (1) classical test theory and (2) components of variance analysis of variance. To these, there is added a dash of Bayesian spirit. Classical test theory is certainly a central feature of validity generalization and for two distinct reasons. First, classical test theory is a simple additive random variable structure of true score and error variable, and it is claimed that the validity generalization model is closely akin to the classical test theory model (Hunter et al., 1982, p. 42). Second, classical test theory provides a conceptual interpretation as to how test and criterion reliability or measurement error can be conceptualized. Because of the central role of classical test theory to validity generalization, the features of classical test theory that are germane for validity generalization will be developed next.

2.2 Classical Test Theory

Assume each person has a fixed true score (i.e., a correct value) on an unobserved random variable T_x. Imagine that T_x is perturbed by random variable E_x so that observations are made on a random variable X_i at two points in time, $i = 1, 2$, with X_i defined by

$$X_i = T_x + E_{xi}, \qquad i = 1, 2, \tag{2.2.1}$$

where T_x, E_{x1}, and E_{x2} are all independent random variables with $\mathscr{E}(E_{xi}) = 0$, and $\mathscr{V}(E_{x1}) = \mathscr{V}(E_{x2})$, and where $\mathscr{E}(W)$ and $\mathscr{V}(W)$ denote expectation and variance of W. The independence assumption of E and T can be relaxed, but their independence is often plausible and will be assumed here. Because observations are made only on the random X_i, Equation (2.2.1) is a simple example of a latent variables model (e.g., Everitt, 1984).

The correlation between X_1 and X_2 is denoted $\rho_{x_1x_2}$, and the correlation between X_i and T_x is denoted ρ_{tx}. Then, a straightforward calculation (e.g., Hogg & Craig, 1978) gives

$$\rho_{xx} \equiv \rho_{x_1x_2} = \rho_{tx}^2 = \mathscr{V}(T_x)/[\mathscr{V}(T_x) + \mathscr{V}(E_{xi})], \tag{2.2.2}$$

where ρ_{xx} is the reliability coefficient. It is always nonnegative and is one when $\mathscr{V}(E_{xi}) = 0$.

In the present context, the X_1 and X_2 represent a job test given twice. In precisely the same way, let $Y_j = T_y + E_{yj}$, $j = 1, 2$, denote the criterion job performance variable. Then, with the same structural assumptions, the reliability of Y is $\rho_{yy} = \rho_{ty}^2$.

The reason for specifying two X variables and two Y variables is to make clear the conceptual interpretation of test and criterion reliability; of course, in most practical test validation settings, there will not likely be a second administration of a test nor the possibility of a second assessment of the criterion variable. There are other methods of assessing reliability, or reliabilities might be obtained from other settings. The key issue, however, concerns the correlation between X and Y.

Focus on the prediction problem from a correlational perspective and consider the correlation between X_1 (or X_2) and Y_1 (or Y_2). Let the errors associated with X_1, that is, E_{x1} and with Y_1, that is, E_{y1} be independent. Then, denote the correlation between X_1 and Y_1 as ρ_{xy}. However, this correlation may be written in terms of the latent correlational parameters as

$$\rho_{xy} = \rho_{t_x t_y}(\rho_{xx}\rho_{yy})^{1/2}, \tag{2.2.3}$$

where $\rho_{t_x t_y}$ is the correlation between the predicted and criterion true score variables, T_x and T_y.

Thus, the X and Y correlation is expressible as a product structure of true score correlations and the square root of the product of the reliability coefficients. This model has been termed the three-component or three-factor model. From a model perspective, one essential goal of validity generalization is to correct upward ρ_{xy} to obtain the true validity correlation, $\rho_{t_x t_y}$, by

$$\rho_{t_x t_y} = \rho_{xy}/(\rho_{xx}\rho_{yy})^{1/2}. \tag{2.2.4}$$

Note that no distributional assumptions need be made for any of the above development, so the above results are very general (only the existence of moments is assumed).

In a typical employment setting, the selection of applicants is based (at least in part) on applicant scores on the test variable X_1. A simple selection model views selection as equivalent to a truncation from below on the X_1 variable, that is, applicants are selected if they have critical scores above some $X_1 = x_{1c}$, for example. In the testing literature, the result of this top down selection procedure is called "restriction of range," but its precise meaning from the perspective of a probability model is often not clear.

If it is assumed that the population model is thus, in reality, the correlation between Y_1 and X_{1c}, where X_{1c} denotes X_1 truncated at from below (i.e., its lower tail is chopped off) at point x_{1c}, under the further assumption that X_1 and Y_1 are jointly normal, Aitkin (1964) provides a correction for selectivity that results in a further correction to the right side of Equation (2.2.3). Let ρ_c be Aitkin's correction (the radical expression in his unnumbered equation, 1964, p. 268). Then, the four-factor equation becomes

$$\rho_{xy} = \rho_{t_x t_y}(\rho_{xx}\rho_{yy}\rho_c)^{1/2}. \tag{2.2.5}$$

Observe that this final correction does not fall out neatly from the foregoing thematic development. Also, it might be wondered if in fact this last correction should be viewed as functionally independent of the other corrections; furthermore, applicant selection may not be as simple as it is portrayed here. While these are important issues, they will not be considered at this point. They are not central to the development. They will however be considered in Chapter 8. Equation (2.2.5) is, however, the four-factor model underlying validity generalization, although it does not include all the corrections that are sometimes suggested.

Of course, Equation (2.2.5) must be written in suitable random variable form since observations are on random variables, never parameters. To anticipate what follows, it is the random variable analogue of $\rho_{t_x t_y}$ that is of central interest. How this structure is cast into a random framework is considered next.

Before doing so however, simplify the notation by letting $\rho_{xy} = \rho$, $\rho_{xx} = \alpha^2$, $\rho_{yy} = \beta^2$, $\rho_c = \delta^2$ and $\rho_{t_x t_y} = \rho'$, where ρ' is the main parameter of interest, the true validity. So, Equation (2.2.3) becomes

$$\rho = \rho' \alpha \beta, \tag{2.2.6}$$

and if four factors are considered, Equation (2.2.5) becomes

$$\rho = \rho' \alpha \beta \delta. \tag{2.2.7}$$

Of course, if unreliability and range restriction are negligible, then $\rho = \rho'$. To anticipate somewhat, Equation (2.2.7) may be viewed as specific values of the random variable from $P = P'ABD$.

2.3 A Validity Generalization Random Model

The easiest way to proceed is simply to define a validity generalization random model, particularize it in certain ways, and then examine its properties to see what characteristics are forced by the assumed structure. Thus, define the validity generalization model as

$$R = P + E, \qquad -1 \le R \le 1. \tag{2.3.1}$$

Let R, P (read rho), and E be random variables that may be regarded as continuous variables although this assumption need not be made for most of the development below. R takes on values r. The interpretation in mind is that R denotes the raw validity coefficients of validation studies. Only observations on R are possible, so Equation (2.3.1) is another example of a latent variables model.

P is a random variable that takes on values ρ. For most of what follows P may have any structure desired. For example, it may be a product structure or an additive composition. Assume P is in the interval $-1 < P < 1$. However, there is nothing special about this particular interval; it could be shorter.

To anticipate somewhat, if P is a product structure with $P = P'ABD$, with P', A, B, and D random variables, P' will be the random variable of true validities, with specific values ρ'; the general goal of validity generalization is to make statements about P', particularily statements about $\mathcal{V}(P')$, the variance of P'.

In Equation (2.3.1), E is an error random variable with $\mathscr{E}(E) = 0$. Let E be distributed on, at this point, some unknown interval; more will be said about E later.

Next, consider a special case of Equation (2.3.1) that is, in fact, a more familiar situation. Envision P as fixed at some value ρ. Such a simplification would be plausible if the focus were on identically replicated validation studies or, less stringently perhaps, when closely related job settings use the same test and criterion variables, so it is sensible to assume P is fixed at some ρ. Denote E and R when P is fixed at $P = \rho$ as R_ρ and E_ρ. If, for example, the test and criterion variables are bivariate normal, with correlation parameter ρ, then R_ρ would have the distribution associated with the sample correlation coefficient for this fixed ρ.

It is important to examine an issue related to this special case of Equation (2.3.1) that may seem a minor matter but turns out to be critical for determining the correlation between P and E in the model of Equation (2.3.1).

Whatever the joint distribution of the test and criterion variables and the associated distribution of R, the distribution of R does not, in general, for a fixed ρ, have expectation ρ (Gayen, 1951). That is, $\mathscr{E}(R_\rho) \neq \rho$, but $\mathscr{E}(R_\rho) = \rho + \Delta_\rho$, where Δ_ρ is a small correction that may depend on ρ; said differently, R_ρ is not an unbiased estimate of ρ. For example, in the bivariate normal case, if $\rho = .5$, with sample size 30, then $\mathscr{E}(R_\rho) = .4935 = \rho + \Delta_\rho$, so $\Delta_\rho = -.0065$.

As a consequence of the fact that R_ρ is not unbiased, there are two ways a model for this special case of Equation (2.3.1) may be defined: the model may be defined so that $R_\rho - \mathscr{E}(R_\rho) = E_\rho$ or, equivalently,

$$R_\rho = \rho + \Delta_\rho + E_\rho, \tag{2.3.2}$$

then $\mathscr{E}(E_\rho) = 0$, for all ρ, and $\mathscr{V}(R_\rho) = \mathscr{V}(E_\rho)$. Alternatively, the model could be defined as

$$R_\rho = \rho + E_\rho^*. \tag{2.3.3}$$

If so, then $\mathscr{E}(R_\rho) = \rho + \Delta_\rho = \mathscr{E}(\rho) + \mathscr{E}(E_\rho^*) = \rho + \mathscr{E}(E_\rho^*)$. Evidently, $\mathscr{E}(E_\rho^*) = \Delta_\rho$. Thus, in Equation (2.3.2), E_ρ is an error variable with mean zero for all ρ, while in Equation (2.3.3), E_ρ^* is an error variable but without mean zero for all ρ. In practical work, the difference between the equations is trivial. Conceptually, the difference is important, as will be seen.

The conceptual difference between Equation (2.3.1) and Equations (2.3.2) and (2.3.3) is also important, and this difference may have lead to confusion in the literature. For example, Callender and Osborn (1980) have written $r_{x^*y^*} = \rho_{x^*y^*} + e$, which is their Equation 3. That $\rho_{x^*y^*}$ is sometimes viewed as a fixed constant is made clear by their comment: "with actual samples there will be some difference between $\rho_{x^*y^*}$ and the correlation $r_{x^*y^*}$ that is computed on the data. This is due to chance sampling error, and we represent it by an additive variable" (1980, p. 548). However, only a few lines later they wrote, as their Equation 8 (which is identical in their notation to), $\mathscr{V}(\rho_{x^*y^*}) + \mathscr{V}(e) =$

$\mathscr{V}(r_{x^*y^*})$; apparently, $\rho_{x^*y^*}$ became a random variable and uncorrelated with e. What they apparently wished to do was to make a distinction between what might be termed the unconditional model of Equation (2.3.1) and the conditional models of Equations (2.3.2) and (2.3.3). Their Equation 3 is similar to Equations (2.3.2) or (2.3.3), but they have no clear unconditional analogue to Equation (2.3.1).

With just this much machinery, some interesting and important properties of the validity generalization model are forced.

2.4 The Joint Space of P and E

Considerable knowledge about the validity generalization model can be obtained by specifying the joint space of P and E, that is, the space in which the pairs of values of P and E [i.e., (ρ, e)] are distributed.

Determining the joint space of P and E is easy. First, consider the space of R_ρ in Equation (2.3.2) [or Equation (2.3.3); the minor difference between them may be ignored for most of what follows, and Δ_ρ can be set to zero]. The critical fact to note is that R_ρ is bounded, that is, $-1 \leq R_\rho \leq 1$. If so, then $-1 \leq \rho + E_\rho \leq 1$ and equivalently $-(1 + \rho) \leq E_\rho \leq 1 - \rho$. Hence, regardless of the distribution of E_ρ, its domain is dependent on ρ. If $\rho = .5$, then $E_{\rho=.5}$ takes on values from -1.5 to $.5$. If $\rho = .8$, $-1.8 \leq E_{\rho=.8} \leq .2$. Thus, for each value of ρ, there must be a different distribution of E_ρ. This result follows from the fact that two random variables have the same distribution if and only if they have the same (cumulative) distribution function (e.g., Clarke, 1975, p. 49). If random variables are distributed on different intervals, as in these examples, it follows immediately they cannot have the same distribution function (Note 2.1).

By considering the range of values of ρ, $-1 < P < 1$, the joint space of P and E is easily graphed. It is the parallelogram in Figure 2.1. Precisely, this parallelogram is the joint space for all coordinate pairs of points ρ and e. Denote this space as $\mathscr{P} = \{(\rho, e): -1 < \rho < 1 \text{ and } -(1 + \rho) \leq e \leq 1 - \rho\}$.

The joint distribution of P and E over $\mathscr{P}$ is denoted here as $f(\rho, e)$ (regardless of what it might be) with $f(\rho, e) = 0$ outside of $\mathscr{P}$, in Figure 2.1. The edges of the rectangle in Figure 2.1 denote the spaces of the marginal distributions of

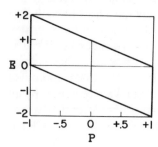

FIGURE 2.1. The parallelogram joint space $\mathscr{P}$ of P and E under the model $R = P + E$. Please see the text for details.

E and P with $-1 < P < 1$, and $-2 < E < 2$, a peculiar range of values for an error distribution. Let the marginal densities of P and E be denoted $f_1(\rho)$ and $f_2(e)$, respectively. Vertical slices inside the parallelogram would denote the spaces (vertical lines) for the distributions of E_ρ with densities $f(e \mid \rho)$, while horizontal slices denote the spaces for the conditional distribution of P given $E = e$.

This joint P and E space is certainly unusual. It is clear from Figure 2.1 that if $\mathscr{E}(E_\rho) = 0$ [which means the model satisfies Equation (2.3.2)], the distribution of each E_ρ cannot be a symmetrical distribution about zero except for $\rho = 0$. Thus, it must be that the conditional error distributions are asymmetric about zero.

2.5 P and E Are Dependent Variables

Once the joint space for P and E has been determined, then the matter of determining whether E and P are dependent or not is trivial. By looking at Figure 2.1, it may be concluded that E and P are dependent random variables. The reason is simple enough. Unless bounded random variables have a distribution on a rectangle, they must be dependent; because P and E are distributed on a parallelogram, they are dependent.

Theorem 2.1 (Simplified). *P and E are not distributed on a rectangle. Therefore, they are dependent random variables.* Please see Note 2.2 for the argument.

This result is very general. Note importantly it does not depend on specifying $f(e, \rho)$, the unknown joint probability distribution of E and P over the space $\mathscr{P}$. The theorem applies not only to the special case of the parallelogram space of Figure 2.1 but to all other nonrectangular spaces, such as circles and ellipsoids.

Given the inherent bounded structure of correlation coefficients, and consequently the boundedness of models of the form of Equation (2.3.1), it does not appear possible to construct a validity generalization model of the form of Equation (2.3.1) with P and E independent.

2.6 The Distribution of R, $c(r)$

Stating that P and E are dependent is an important result, and it has critical implications for attempts to develop estimation procedures for models of the form of Equation (2.3.1). Remember that a main goal of validity generalization—infact the goal—is to make statements about P, precisely the mean and variance of P through knowledge of the mean and variance of R. To make any progress on the problem, it seems essential that the distribution of R be specified. Stating that P and E are dependent virtually precludes this possibility. To see why, consider the procedure necessary for obtaining the distribution of R, $c(r)$ based on a model of the form of Equation (2.3.1).

First, the joint distribution $f(e, \rho)$ over the space $\mathcal{P}$ in Figure 2.1 must be specified. Just what this distribution might be is by no means clear. Second, with $f(e, \rho)$ specified, the distribution function of R, $C(r)$ could be obtained by integrating over $\mathcal{P}$, in Figure 2.1 for all (ρ, e) in $\mathcal{P}$; thus,

$$C(r) = P(R \leqq r) = \iint\limits_{e+\rho \leqq r} f(\rho, e) \, d\rho \, de. \qquad (2.6.1)$$

That is, the integration is over all pairs (ρ, e) that have sums less than or equal to some specified r as r ranges from -1 to 1. This is not an easy numerical integration in the plane. However, the task would be much easier if P and E were independent, because $f(\rho, e)$ could be factored into $f(\rho, e) = f_1(\rho)f_2(e)$ (Hogg & Craig, 1978, Theorem 1, p. 81) and two separate univariate numerical integrations could be achieved, a far less formidable task. Since E and P are dependent, $f(\rho, e)$ cannot be so factored. Furthermore, once $C(r)$, the distribution function of R, has been obtained, its derivative $C'(r)$ gives the density of R, $c(r)$. Again, numerical methods would doubtlessly be needed to obtain $c(r)$. Thus, the entire process would be so computer intensive, that although there appears nothing in principle to prevent such a development, it would likely be so complex it would never be used even if it were developed. Note importantly that none of the above problems speak to the issue of whether Equation (2.3.1) forms an adequate conceptual approach to the general problem of interpreting Figure 1.1.

Of course, knowing $c(r)$ does not reveal the mean or variance for R, which would most likely be a more difficult numerical integration than Equation (2.6.1). Finally, keep in mind that all this development has very little to say about how the parameters of the latent structure of Equation (2.3.1) are to be estimated should the distribution $c(r)$ be obtained. That is another quite separate problem, but it would depend on the distribution $c(r)$.

From one perspective, it seems surprising, given the central importance of Equation (2.6.1) to the validity generalization model, estimation, and fitting procedures, there has apparently been no work on the problem. From another perspective, given the difficulty of fully specifying Equation (2.6.1), it appears to be a hopelessly complex problem.

Note that $c(r)$ from Equation (2.6.1) might form the basis for a specified sampling model that would provide the foundation for specifying the parent distribution for empirical validity coefficients, such as in Figure 1.1. Certainly, any rigorous model for validity generalization must specify $c(r)$.

2.7 Are P and E Correlated?

There has been considerable concern about whether P and E in Equation (2.3.1) are correlated or not (e.g., James, Demaree, & Mulaik, 1986). This concern is well placed. Life is much easier if one could write $\mathcal{V}(R) =$

$\mathscr{V}(P) + \mathscr{V}(E)$ rather   than $\mathscr{V}(R) = \mathscr{V}(P) + 2\,\mathrm{cov}(P, E) + \mathscr{V}(E)$,   where $\mathrm{cov}(P, E)$ is the covariance of P and E.

There are a couple of ways to get rid of the P and E covariance term. One possibility is to assume E and P are independent; in this case, the covariance term vanishes because independence implies zero covariance or equivalently zero correlation. Section 2.5 makes it clear this assumption, which has been made (e.g., Callender & Osburn, 1980, p. 548) and which Burke (1984) regards as essentially a working assumption, is wrong. Actually, independence is not needed, so, for example, Callender and Osburn (1980) state a stronger condition than they need: only an uncorrelated condition is required. Thus, if P and E are zero correlated, then the covariance is zero.

Another approach to the troublesome covariance term is to hope it is negligible and proceed as if it were so. From the standpoint of practical work over the last decade, hope seems to have been the reed to lean on. It turns out that it may be a sufficiently sturdy reed. Indeed, it is possible to state a condition that will guarantee P and E will be uncorrelated. Before doing so, however, it is instructive to consider two examples.

EXAMPLE 2.7.1. Suppose $f(\rho, e) = \frac{1}{4}$ over the set $\mathscr{P}$ in Figure 2.1. Pictorially, this density would be a parallelepiped. The marginal density of P is $f_1(\rho) = \frac{1}{2}$, $-1 < \rho < 1$. E has marginal density $f_2(e)$, which is an isoceles triangle with apex at zero, $f_2(0) = \frac{1}{2}$, $-2 < e < 2$. Figure 2.2 shows this joint density and the marginal distributions. A straightforward calculation gives the correlation between E and P as $-.71 = -1/\sqrt{2}$, quite a sizable negative correlation. Of course, this joint density is not a plausible candidate for the distribution of $f(\rho, e)$. Observe that in this simple example, the integral defined by Equation (2.6.1) is easily obtained. It is $C(r) = r/2 + \frac{1}{2}$ and $c(r) = \frac{1}{2}$, $-1 \leqq r \leqq 1$.

EXAMPLE 2.7.2. With considerably more effort, it is possible to be more realistic. Callender and Osborn (1980) proposed what they called a normal distribution of true validities, which is their Table 3. Although their normal

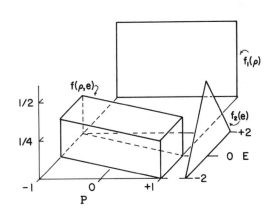

FIGURE 2.2. The joint distribution $f(\rho, e) = \frac{1}{4}$ over the region $\mathscr{P}$, and the marginal densities of P and E, $f_1(\rho)$, and $f_2(e)$.

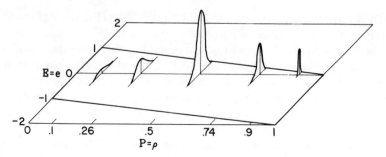

FIGURE 2.3. A portion of the joint space $\mathscr{P}$ of P and E when the density $f(e, \rho)$ over the space is determined by Callender and Osborn's (1980) distribution of P, and R_ρ is g distributed. These slices, which are elements of the joint space, would be conditional densities $f(e \mid \rho)$ if properly normalized.

distribution is a finite distribution and does not even cover the interval 0 to 1, it may still be a crude starting point for determining a distribution of $f(e, \rho)$. Let their normal distribution define $f_1(\rho)$. Next, assume the distribution of R_ρ, that is, the conditional distribution of R when P is fixed at $P = \rho$, is g distributed, where g distributed denotes the distribution of R that arises when the variables involved are bivariate normal (see Note 2.3). The distribution of $R_\rho - \rho = E_\rho^*$ [from Equation (2.3.3)], is a straightforward change of variables calculation (cf. Hogg & Craig, 1978); the conditional distribution of $f(e \mid \rho)$ was obtained from the approximate distribution of g given by Soper, Young, Cave, Lee, and Pearson (1917) as their Equation lii. Once $f_1(\rho)$ and $f(e \mid \rho)$ are known, then $f(\rho, e) = f_1(\rho)f(e \mid \rho)$ may be easily determined.

The graph of $f(\rho, e)$, for $n = 50$, is shown for selected values of P in the restricted space of Figure 2.3. Since $f_1(\rho)$ is discrete, $f(\rho, e)$ is discrete over P and continuous over E. Thus, the slices pictured in Figure 2.3 are elements of this joint space. A straightforward numerical integration can be used to determine the covariance of E and P. Looking at Figure 2.3, the correlation appears to be approximately zero. The covariance is about $-.02$; thus, the correlation between P and E would most likely be neglible.

These two numerical examples illustrate the dependency of the P and E correlation on the distribution $f(\rho, e)$. They also suggest that under some distributions the correlation may be negligible, although possibly slightly negative. James et al. (1986) also concluded, from a different perspective, that P and E would be negatively correlated. However, it is possible to state a largely distribution-free result, which is a sufficient condition to guarantee that P and E will be uncorrelated. Precisely, P and E are uncorrelated if Equation (2.3.2) holds, that is, if $\mathscr{E}(E \mid \rho) = 0$ for all ρ.

Theorem 2.2 (Simplified). *If $\mathscr{E}(E \mid \rho) = 0$ for all ρ, then P and E are uncorrelated.* Please see Note 2.4 for the argument.

Thus, the model of Equation (2.3.1) forms the basis for an interesting and infrequently observed conceptual result, particularily for a presumably real-world model. P and E are clearly *dependent* but will be *uncorrelated* if the model satisfies Equation (2.3.2). Note that the numerical calculation based on Callender and Osburn's (1980) $f_1(\rho)$ distribution used Equation (2.3.3) to define the model. Thus, the slight negative correlation obtained would have been suggested given knowledge of Theorem 2.2.

The issue of whether P and E are uncorrelated or not is not simply a matter of blithely declaring that Equation (2.3.2) is the model. The issue depends on the distribution $f(\rho, e)$. Once $f(\rho, e)$ has been selected, Theorem 2.2 states a sufficient condition for P and E to be uncorrelated.

Note in Example 2.7.1 when a parallelopiped was considered as a density over $\mathcal{P}$ Equation (2.3.2) does not hold. In this example, $\mathscr{E}(E) = 0$, but $\mathscr{E}(E \mid \rho) \neq 0$ except for $\rho = 0$, which illustrates that simply assuming that the unconditional error distribution E has mean zero is not sufficient to force an uncorrelated condition for P and E. Thus, $\mathscr{E}(E) = 0$ does not imply that $\mathscr{E}(E_\rho) = \mathscr{E}(E \mid \rho) = 0$. On the other hand, $\mathscr{E}(E_\rho) = 0$ for all ρ does imply $\mathscr{E}(E) = 0$.

If $f(\rho, e)$ is suitably derived from the g distribution of R_ρ, then it appears P and E will be, at least for practical purposes, essentially uncorrelated.

2.8 Identifiability

Constructing a probability model is one matter. Specifying a model for which the unknown parameters may be estimated in data may be quite another. The problem for consideration is the problem of parameter identifiability, which is best introduced with examples.

EXAMPLE 2.8.1. Suppose X is a normal random variable with variance σ^2 and parameters μ_1 and μ_2, where $\mu_1 - \mu_2 = \theta$ is the mean parameter. It is desired to estimate μ_1 and μ_2 from observations of X. However, it is easily seen that there are an infinite number of μ_1 and μ_2 pairs with the same difference θ. Consequently, μ_1 and μ_2 cannot be estimated and thus are not identifiable.

Intuitively, for a parameter to be identifiable, changes in the value of the parameter must result in changes in the functional form of the distribution of the random variable in question. Because different values of μ_1 and μ_2 result in the same θ and, consequently, the same functional form of the normal density of X, μ_1 and μ_2 are said to be not identifiable by X. Alternatively, the mean parameter θ is identifiable by X because changes in θ alter the distribution of X. If $X = V - W$, where V and W are normal with means μ_1 and μ_2, then the difference is identifiable if observations on V and W are possible, and μ_1 and μ_2 are identifiable by V and W.

Thus, if two distributions are identical but have a pair of parameter values that are different, the parameters are not identifiable (see Basu, 1983, p. 3, for a more formal definition of nonidentifiability), and the distributions are said to be *observationally equivalent* because they have the same probability structure (Schmidt, 1983, p. 10). Specifically, if Y is normal with variance σ^2 and mean $\theta = \mu_1' - \mu_2'$, then X and Y have the same distribution and are observationally equivalent with respect to the μ parameters.

Because observationally equivalent distributions with respect to the μ parameters possess no information about the μ parameters, clearly random samples from the same distribution as X and Y cannot provide such information either.

This example clearly indicates that logically prior to any consideration of estimation it must be able to be asserted that the parameters or constants that are to be estimated are identifiable. If they are not identifiable, estimation is meaningless.

That there is a serious identification problem for a general multifactor validity generalization model, such as Equations (2.2.6) and (2.2.7), is easily illustrated.

EXAMPLE 2.8.2. Let R be a g-distributed correlation coefficient. Then, the density of R depends only on ρ, the population correlation coefficient, and n, the sample size. Suppose ρ is fixed at $\rho = \rho'\alpha\beta$, the three-factor model of Equation (2.2.6). Clearly, there are many values of ρ', α, and β with product ρ; consequently, these parameters are not identifiable and hence cannot be estimated. Thus, although the parameterized model may seem plausible, the latent parameters cannot be estimated. Of course, if these parameters are known, the problem is very different. Suppose $\alpha\beta$ is known and the observed $R = r$, where R estimates $\rho'\alpha\beta$. Then, a natural estimate of ρ' to consider is $r/\alpha\beta$ (but see Section 8.1.1). In the general setting, clearly the parameters of the product structure are unknown. Therefore, there is no well-defined statistical estimation problem, at least for the general factor model.

This example has important consequences for the validity generalization procedures (cf. Burke, 1984), which attempt estimation of the nonidentifiable $\mathscr{V}(P')$ in the multifactor model. Just what the many and widely used estimation procedures are in fact estimating will be considered later. For the moment, consider the depth of the problem.

The essence of the problem from an identifiability perspective is therefore the following. Suppose

$$R = P'A, \tag{2.8.1}$$

which may be viewed as Equation (2.3.1), where $P = P'ABD$ but with the simplification that B, D, and E are degenerate random variables with point masses as 1, 1, and 0, respectively.

Let it be assumed that $\mathscr{V}(R)$ is known. Do unique variances of P' and A for a fixed $\mathscr{V}(R)$ exist? If so, then there might be some hope of estimating the variances. If not, then estimation is hopeless, because if there is no unique variance of P' for a fixed variance of R in the model, there can be no information in data generated under that model that would lead to a unique estimate of $\mathscr{V}(P')$.

EXAMPLE 2.8.3. Suppose $R = P'A$ has distribution

$$w(r) = w(\rho'a) = \tfrac{1}{4} \quad \text{if} \quad \rho'a = .10, .12, .125, .15$$
$$= 0 \quad \text{otherwise.}$$

Thus,

$$\mathscr{V}(R) = \mathscr{V}(P'A) = .00032$$
$$= \tfrac{1}{4}(.10^2 + .12^2 + .125^2 + .15^2) - [\tfrac{1}{4}(.10 + .12 + .125 + .15)]^2.$$

Let P' have the distribution

$$u(\rho') = \tfrac{1}{2} \quad \text{if} \quad \rho' = .4, .5$$
$$= 0 \quad \text{otherwise,}$$

and A have the distribution

$$v(a) = \tfrac{1}{2} \quad \text{if} \quad a = .25, .3$$
$$= 0 \quad \text{otherwise.}$$

The distribution of the product $P'A = R$ under independence is precisely the distribution $w(r) = w(\rho'a)$, and the $\mathscr{V}(P') = .0025$. With the example just given listed first, consider 5 possible distributions of P' and A and their product distribution $R = P'A$. All have common variance $\mathscr{V}(R) = .00032$ and different $\mathscr{V}(P')$ listed in Table 2.1.

Assuming P' and A are independent, and with equal outcome probabilities, the product of each $P'A$ pair has a distribution that is precisely the distribution $w(r) = w(\rho'\alpha)$. Thus, all 5 pairs of random variables $R = P'A$ have observationally equivalent distributions. Yet, as is clear also, the variance of P', $\mathscr{V}(P')$,

TABLE 2.1. Five observationally equivalent distributions of $R = P'A$ with common $\mathscr{V}(R)$ and different $\mathscr{V}(P')$.

Distribution of A				Distribution of P'				$\mathscr{V}(P')$
.25	.3			.4	.5			.0025
.5	.6			.2	.25			.00063
1.0	1.0			.10	.12	.125	.15	.00032
.9	.75			.133	.166			.00028
.2	.24	.25	.30	.5	.5			0

Note. For each distinct distribution of A and P', the outcome probabilities are equal. The distribution of $R = P'A$ when each A and P' are independent is identical and $\mathscr{V}(R) = .00032$.

is different in each case. It can exceed the variance of R, it can be less than the variance of R, or it can be zero.

These counterexamples show that in general there is no unique $\mathscr{V}(P')$ for a fixed $\mathscr{V}(R)$, which was the central point to be made.

Note that it has not been shown that $\mathscr{V}(P')$ can never be uniquely specified. There exists no general theorem guaranteeing when model parameters are or are not indentifiable. Each situation must be approached anew.

Let there be no doubt that the identifiability problem is absolutely crucial for validity generalization and until the issue is resolved, there is no well-defined statistical problem. Furthermore, let it be emphasized that the importance of parameter identifiability is simply not in dispute. For example, Basu says, "Before any inferential procedure can be developed, one needs to assert that the unknown parameters are identified." (1983, p. 2). Schmidt states that "If two parameter values imply the same distribution of data, no observed data can distinguish between them, and there is no point in attempting to estimate those parameters. Thus consideration of identification is a question that logically precedes estimation." (1983, p. 13).

Nonidentifiability is a problem in many models. The general factor analysis model and the LISREL model are examples (cf., Everitt, 1984). One solution to the problem is to try to constrain the model in certain ways so that the constants to be estimated may be identified. There is, however, no general strategy for achieving model identifiability. But, it is a crucial, nontrivial problem to state conditions that make the constants of the validity generalization multifactor model identifiable.

This section has been included in anticipation of considering the validity generalization multifactor estimation procedures to be discussed in Section 3.1.4. Whether these unorthodox estimation methods serve as examples of attempts to estimate unidentified parameters or not may be a point of contention. What is clear, however, is that one goal of the these procedures is to estimate the variability of the ρ' when the ρ' are embedded in a product of the form $\rho'\alpha\beta$ and where α and β are almost always unknown.

3
Estimation

3.1 Introduction

If the conclusions of the previous chapter are accepted, it might be wondered why estimation would be attempted under such an analytically intractible and generally unidentified model. The explanation may be simple. In effect, there has been estimation without a model. Recall that Schmidt and Hunter (1977) sketched only an idea: no model was presented, but they nonetheless provided an estimation scheme. This same trend has continued. In fact, there are a variety of related, generally complex, estimation schemes (cf. Burke, 1984), and almost all of them are computer intensive—cut loose as it were from any formal model to guide interpretation. The consequences of this state of affairs are easily predicted: the problem both practically and conceptually is to make sense out of the estimates provided.

Given the widespread use of validity generalization procedures, it is necessary to consider what the estimates may mean. The best that can be done is to consider them on their own terms, examining them with standard tools. Fortunately, this is not difficult because all estimation schemes in common use make use of two sample quantities that will be defined and studied in the following sections. Let the reader be forewarned, however, that given the results in Chapter 2, it should not be surprising to discover that the estimators may have strange and unconventional properties.

Before examining the validity generalization estimators, however, it will be useful to put estimation methods into a broader perspective so that the methods used in validity generalization may be viewed against a backdrop of modern statistical inference.

3.2 Lehmann's Classification

E.L. Lehmann (1983) views estimation methods as classifiable into one of three general approaches.

3.2.1 DATA ANALYSIS

Here, data are analyzed with few, if any, assumptions. The computation of ordinary means and medians for summary purposes would be an example. Least squares procedures as well as robust methods would also be classified here. Inferential methods would for the most part not be regarded as data analytical methods because inference presumes some model structure.

3.2.2 CLASSICAL INFERENCE AND DECISION THEORY

The crucial distinguishing feature of this approach, as well as the next approach, is that the observations are assumed to have a specified joint probability distribution. For example, if $R_1, \ldots, R_k$ were a random sample from the distribution $c(r)$, of Equation (2.6.1), then the joint probability density would be given by

$$\prod_{i=1}^{k} c(r_i), \tag{3.2.1}$$

and parameter estimation would focus on the joint density. However, $c(r)$ has never been specified. Consequently, validity generalization methods are not classical inference methods.

3.2.3 BAYESIAN ANALYSIS

Bayesian methods also require the specification of a joint probability structure just as in classical methods. In addition, however, the parameters are not assumed to have fixed values as in classical methods but rather they are assumed to follow a distribution of their own.

Validity generalization methods thus fall outside the domain of both classical and Bayesian methods, at least as viewed by Lehmann. While it has been claimed even in the titles of works that validity generalization methods were Bayesian methods (Pearlman, 1982; Schmidt, Hunter, Pearlman & Shane, 1979), these writers have used the language of Bayesian methods but not the procedures. The relationship between validity generalization and Bayesian methods will be explored in Section 4.5.

3.3 Sample Data

Most familiar statistical procedures assume random sampling, which means that each random variable in the sample is independent and identically distributed (iid). That structure is inappropriate in the validity generalization setting.

Even if $R_i, i = 1, \ldots, k$ were independent correlation coefficients they would not be identically distributed. This is because the distribution of each R_i depends on sample size, n_i, which is a parameter of the distribution of R_i, and sample size varies substantially across validation studies. In published validity distributions (e.g., Carruthers, 1950; Schmidt & Hunter, 1978), the largest sample size may be 5 or 6 times the smallest, and sample sizes may range from 10 to several hundred although the average sample size is about 68 (Lent, Aurbach & Levin, 1971). Thus, among the k R_i, there are at least as many different distributions represented as there are differences in sample size.

Furthermore, examination of these same published validity coefficients suggests that the same subjects may be used in different validation studies, particularily if different validity coefficients are contributed by the same organization. Consequently, it seems likely that in some cases the k R_i are not mutually independent. Both of these facts will influence the interpretation of the sample quantities to be given in the following.

3.4 The Basic Sample Estimates

Because no joint density can be specified, the two main inferential model approaches defined by Lehmann (1983) are not possibilities. About the only possibility is a straightforward partitioning of sums of squares; consider where this leads.

From Equation (2.3.1) let there be $R_i = P_i + E_i$, $i = 1, \ldots, k$. Define $\bar{R} = \sum R_i/k$, and similarly define $\bar{P}$ and $\bar{E}$. Then, a simple expansion gives the following sample quantities:

$$S_R^2 = S_P^2 + S_{PE} + S_E^2 \tag{3.4.1}$$

where

$$(k - 1)S_R^2 = \sum (R_i - \bar{R})^2, \tag{3.4.2}$$

$$(k - 1)S_P^2 = \sum (P_i - \bar{P})^2, \tag{3.4.3}$$

$$(k - 1)S_{PE} = 2 \sum (P_i - \bar{P})(E_i - \bar{E}), \tag{3.4.4}$$

and

$$(k - 1)S_E^2 = \sum (E_i - \bar{E})^2. \tag{3.4.5}$$

Observations are made only on the R_i, so only S_R^2 is computable. The cross product sample covariance term S_{PE} is assumed to be neglible or zero. Given Theorem 2.2, such an assumption may not be too misleading. Because S_E^2 is not computable, it is replaced with S_*^2, where

$$S_*^2 = k^{-1} \sum (n_i - 1)^{-1}(1 - R_i^2)^2. \tag{3.4.6}$$

The replacement of the least squares estimate S_E^2 with S_*^2 is not to be taken lightly. The form of S_*^2 is very different from the other estimators in Equation (3.4.1), all of which are distribution-free estimators. The properties of S_*^2 will be studied later. However, some conceptual justification for using S_*^2 as a replacement for S_E^2 can be obtained through the following development.

Consider the conditional distribution of R given $P = \rho$ in Equation (2.3.2). Then, $\mathscr{V}(R_\rho) = \mathscr{V}(\rho + \Delta\rho + E_\rho) = \mathscr{V}(E_\rho) \approx (1 - \rho^2)^2/(n - 1)$, where $(1 - \rho^2)^2/(n - 1)$ is the leading term in the expansion of the variance of R in the bivariate normal case (Patel & Read, 1982, p. 313). Then, replacing ρ with $R_\rho = r_\rho$, an observed value of R_ρ, gives $\hat{\mathscr{V}}(R_\rho) \approx (1 - r_\rho^2)^2/(n - 1)$, which should be viewed as an estimate of an approximation and thus could be a very poor estimate. Now, thinking unconditionally, take each $(1 - r_i^2)^2/(n_i - 1)$ and average over all k of them. This average gives S_*^2. Thus, Equation (3.4.1) is altered so that S_P^2 is defined by

$$S_P^2 = \frac{\sum (R_i - \bar{R})^2}{(k - 1)} - \sum \frac{(1 - R_i^2)^2}{k(n_i - 1)}. \qquad (3.4.7)$$

Then, a point estimate of S_P^2 is obtained from

$$s_P^2 = \frac{\sum (r_i - \bar{r})^2}{(k - 1)} - \sum \frac{(1 - r_i^2)^2}{k(n_i - 1)}, \qquad (3.4.8)$$

where observed validity coefficients are the r_i.

Thus, S_P^2 is intended to reflect an adjustment of S_R^2, the variability among correlation coefficients, by the average variability associated with each correlation. The expression does, at first glance anyway, have a certain intuitive plausibility. However, Equation (3.4.7) shows only a vague resemblance to the partitioning of the sums of squares and cross products in Equation (3.4.1). Note that S_P^2 is not $S_\bar{P}^2$. For instance, it is always true that $S_\bar{P}^2 \geq 0$, but it is not obvious $S_P^2 \geq 0$. Consequently, there is no assurance that the properties of $S_\bar{P}^2$ will apply to S_P^2. Presumably, S_P^2 estimates the variance among the ρ_i population correlation coefficients. These expressions may differ slightly from those used in the literature; sometimes the R_i are sample size weighted.

3.4.1 COMMENT

While the original motivation for attempting to resolve the specificity–generalizability problem from the perspective of what is now termed validity generalization seems unclear, it was likely the realization that the variance of the E_i might be estimated from one function of R_i, namely, $\sum(1 - R_i^2)^2/(n_i - 1)$, while the overall variability of the R_i could be estimated from another function of the same R_i, namely, the sample variance S_R^2, was a key idea in this development.

In any event, the essence of the estimation procedure is Equation (3.4.8), which is sometimes called the bare bones estimate. More complicated procedures simply consider functions of S_R^2 and S_*^2 in an attempt to partition the componential variance further. Because of the importance of these two statistics to all validity generalization procedures, it is important to consider their properties.

3.5 The Estimator $S_P^2 = S_R^2 - S_*^2$

Inspecting Equation (3.4.7) is instructive because it portends difficulties ahead. First, what kind of estimator is it? Certainly, it is not a least squares variance estimator, because all such estimators must be nonnegative. To estimate a variance sensibly, S_P^2 must at least be nonnegative. But, S_P^2 clearly can be negative. A simple example would be to take $r_i = 0$ for all i, then $S_R^2 = 0$, but $S_*^2 = k/\sum(n_i - 1)$ for all k, its maximum possible value. Thus, clearly S_P^2 can be negative, and this example shows that S_P^2 can have negative expectation: let the above values be model parameter values. That S_P^2 can be negative and have negative expectation for plausible distributions of R_i will be clear in Section 3.9. The fact that S_P^2 can have negative expectation sets it apart from other models where negative variance estimations are not infrequently observed, such as random effects analysis of variance. In such models, the estimators are typically unbiased.

Second, the denominators of S_R^2 and S_*^2 in Equation (3.4.7) are strangely different, unlike Equation (3.4.1). This suggests the behavior of S_P^2 may be peculiar as n_i and k change. Third, it can be anticipated that the expectation of S_P^2 will be complicated. That is because while S_R^2 is quadratic in R_i, clearly S_*^2 is quartic in R_1, so whatever $\mathscr{E}(S_P^2)$ might be, it will likely be a function of higher order population moments, and not simply the population means and variances.

3.6 Interpreting Estimators

One standard method of assessing what an estimator estimates is to consider the estimator's expectation. Suppose S_P^2 estimates variability among the ρ_i. Define

$$\xi = (k - 1)^{-1} \sum (\rho_i - k^{-1} \sum \rho_i)^2. \tag{3.6.1}$$

Then define the validity generalization hypothesis H_g as

$$H_g: \rho_1 = \rho_2 = \cdots = \rho_k, \tag{3.6.2}$$

which can be given an equivalent representation:

$$H_g \quad \text{holds if and only if} \quad \xi = 0.$$

Consequently it would be hoped that $\mathscr{E}(S_P^2)$ would equal

$$(k - 1)^{-1} \sum [\mathscr{E}(R_i) - k^{-1} \sum \mathscr{E}(R_i)]^2 = B, \tag{3.6.3}$$

and B would be exactly zero if all $n = n_i$ and approximately zero if the n_i varied and if H_g were true.

Because $\mathscr{E}(S_P^2) = \mathscr{E}(S_R^2 - S_*^2) = \mathscr{E}(S_R^2) - \mathscr{E}(S_*^2)$, it is sufficient to consider the expectations of the 2 quantities S_R^2 and S_*^2 separately.

3.7 The Expectation of S_R^2

There are several expressions for $\mathscr{E}(S_R^2)$. Which expression is appropriate depends on the assumptions imposed on the R_i. By expanding S_R^2 and taking expectations, the following distribution-free results are obtained. If desired $\mathscr{E}(R_i)$ may be roughly approximated by ρ_i and $\mathscr{V}(R_i)$ approximated by $(1 - \rho_i^2)^2/(n_i - 1)$.

The most general expectation is

$$\mathscr{E}(S_R^2) = \frac{\sum [\mathscr{V}(R_i) + \mathscr{E}^2(R_i)]}{k} - \frac{2 \sum\sum_{i<j} \mathscr{E}(R_i R_j)}{k(k-1)}, \tag{3.7.1}$$

which allows for dependency among the R_i. If different R_i involve the same subjects, then Equation (3.7.1) is as simple as the expectation becomes, because it would be implausible to assume that the R_i were always pairwise independent.

If all the R_i are pairwise independent so that $\mathscr{E}(R_i R_j) = \mathscr{E}(R_i)\mathscr{E}(R_j)$, $i \neq j$, then

$$\mathscr{E}(S_R^2) = \frac{\sum \mathscr{V}(R_i)}{k} + \frac{\sum\sum_{i<j} [\mathscr{E}(R_i) - \mathscr{E}(R_j)]^2}{k(k-1)}. \tag{3.7.2}$$

Because

$$k \sum [(\mathscr{E}(R_i) - k^{-1} \sum \mathscr{E}(R_i)]^2 = \sum\sum_{i<j} [\mathscr{E}(R_i) - \mathscr{E}(R_j)]^2,$$

Equation (3.7.2) is identically

$$\mathscr{E}(S_R^2) = A + B, \tag{3.7.3}$$

where A and B are defined by

$$A = \frac{\sum \mathscr{V}(R_i)}{k}$$

and

$$B = \frac{\sum [\mathscr{E}(R_i) - k^{-1} \sum \mathscr{E}(R_i)]^2}{(k-1)}.$$

Note that B is identically Equation (3.6.3).

If, in addition, all R_i have the same mean, then

$$\mathscr{E}(S_R^2) = \frac{\sum \mathscr{V}(R_i)}{k} = A. \tag{3.7.4}$$

Finally, if it is further assumed that all R_i share the same variance $\mathscr{V}(R_i)$, then

$$\mathscr{E}(S_R^2) = \mathscr{V}(R_i), \tag{3.7.5}$$

which is also the familiar unbiased estimator under a random sampling model (in which case all R_i are iid). Clearly, Equation (3.7.5) can never hold in practical cases even if all $\rho_i = \rho$. The observations are not iid because sample size varies.

The weakest plausible assumption would seem to be that all validity coefficients are pairwise independent. If so, then Equation (3.7.3) or (3.7.2) is the appropriate expectation. Then, under the validity generalization hypothesis that all $\rho_i = \rho$, it would be expected that B would be about zero. The problem is to find an estimate of $A = k^{-1} \sum \mathscr{V}(R_i)$, which could be taken as the subtractive correction to S_R^2 so the difference is B in Equation (3.7.3). S_*^2 has been taken as that corrective estimate.

3.8 The Expectation of S_*^2

Because

$$\mathscr{E}(1 - R_i^2)^2 = 1 + \mathscr{E}(R_i^4) - 2\mathscr{E}(R_i^2),$$

or identically, if one prefers,

$$\mathscr{E}(1 - R_i^2)^2 = 1 + \mathscr{E}(R_i^4) - 2\{\mathscr{V}(R_i) + [\mathscr{E}(R_i)]^2\},$$

then

$$\mathscr{E}(S_*^2) = \sum \frac{[1 + \mathscr{E}(R_i^4) - 2\{\mathscr{V}(R_i) + [\mathscr{E}(R_i)]^2\}]}{k(n_i - 1)}$$

$$= \sum \frac{1 + \mathscr{E}(R_i^4) - 2\mathscr{E}(R_i^2)}{k(n_i - 1)}, \tag{3.8.1}$$

which is the practically applicable expectation. If $n_i = n$, and all R_i are iid with the same distribution as R, then

$$\mathscr{E}(S_*^2) = \frac{1 + \mathscr{E}(R^4) - 2\mathscr{E}(R^2)}{n - 1}. \tag{3.8.2}$$

Thus, even in the unrealistic case where the R_i are iid, the expectation has no simple interpretation. Clearly, it does not have the desired expectation A. Even if it were assumed that $\mathscr{E}(R^4)$ were neglible and with $\mathscr{V}(R)$ approximated by $(1 - \rho^2)^2/(n - 1)$, and $\mathscr{E}(R)$ by ρ, then

$$\mathscr{E}(S_*^2) = \frac{(n - 3)[1 - 2\rho^2] - 2\rho^4}{(n - 1)^2}, \tag{3.8.3}$$

which again is not easily interpreted.

3.9 The Expectation of S_P^2

Because expectations of differences are simply differences of the expectations, $\mathscr{E}(S_P^2)$ is simply the difference between the appropriate expressions above.

In the general case, $\mathscr{E}(S_P^2)$ is simply Equation (3.7.1) minus Equation (3.8.1) or

$$\mathscr{E}(S_P^2) = \mathscr{E}(S_R^2 - S_*^2) = \frac{\sum [\mathscr{V}(R_i) + \sum \mathscr{E}^2(R_i)]}{k} - \frac{2 \sum_{i<j} \sum \mathscr{E}(R_i R_j)}{k(k-1)}$$

$$- \sum \frac{1 + \mathscr{E}(R_i^4) - 2\mathscr{E}(R_i^2)}{k(n_i - 1)}. \tag{3.9.1}$$

For the case in which R_i and R_j, $i \neq j$, are independent, Equation (3.9.1) becomes

$$\mathscr{E}(S_P^2) = \frac{\sum \mathscr{V}(R_i)}{k} + \frac{\sum [\mathscr{E}(R_i) - k^{-1} \sum \mathscr{E}(R_i)]^2}{k-1} - \sum \frac{1 + \mathscr{E}(R_i^4) - 2\mathscr{E}(R_i^2)}{k(n_i - 1)} \tag{3.9.2}$$

or

$$\mathscr{E}(S_P^2) = A + B - \sum \frac{1 + \mathscr{E}(R_i^4) - 2\mathscr{E}(R_i^2)}{k(n_i - 1)}.$$

If all the R_i have equal means, then Equation (3.9.2) becomes

$$\mathscr{E}(S_P^2) = A - \sum \frac{1 + \mathscr{E}(R_i^4) - 2\mathscr{E}(R_i^2)}{k(n_i - 1)}, \tag{3.9.3}$$

and under the iid case ($k \geq 2$),

$$\mathscr{E}(S_P^2) = \mathscr{V}(R) - \frac{1 + \mathscr{E}(R^4) - 2\mathscr{E}(R^2)}{n - 1}. \tag{3.9.4}$$

Collectively, these expressions have no simple interpretation. Certainly, there is no unambiguous conceptual interpretation for S_P^2. It is not simply an estimate of the variability among the ρ_i as might be hoped.

3.10 Numerical Evaluation of $\mathscr{E}(S_P^2)$

About the only way to gain further understanding of $\mathscr{E}(S_P^2)$ is by numerical evaluation. Suppose each R_i is g distributed. By using the approximations given by Johnson and Kotz (1970, p. 225) or Ghosh (1966) for the moments of R (ignoring the remainder term) and rewriting these central moments in terms of the moments about zero, $\mathscr{E}(R_i^2)$ and $\mathscr{E}(R_i^4)$ may be given, and Equation (3.9.2) and hence $\mathscr{E}(S_P^2)$ may be evaluated. Note that if the R_i were not independent, $\mathscr{E}(S_P^2)$ could not be numerically evaluated because $\mathscr{E}(R_i R_j)$ in Equations (3.7.1) and (3.9.1) would be unknown.

It is instructive to evaluate $\mathscr{E}(S_P^2)$ from the perspective of a hypothesis testing framework. Recall that H_g is defined by Equation (3.6.2) as

$$H_g: \rho_1 = \rho_2 = \cdots = \rho_k.$$

TABLE 3.1. Approximate expected
values of Equation (3.9.4) with $k \geqq 2$.

ρ	$n = 30$	$n = 50$
0	.00199	.000767
.2	.00176	.000676
.4	.00114	.000442
.6	.000428	.000176
.8	−.0000305	.0000132
.9	−.0000130	−.00000477

Consider a contrasting specificity hypothesis:

H_s: Not all the ρ_i are identical,

which is equivalent to $\xi > 0$, where ξ is defined in Equation (3.6.1). From the perspective of validity generalization procedures, H_s is regarded as the null and H_g is the alternative: "If the remaining variance is zero or near zero, the hypothesis of situation specificity is rejected" (Schmidt et al., 1979, p. 261).

First, consider the expected values of S_P^2 under selected H_g. Table 3.1 evaluates Equation (3.9.4) for two sample sizes and selected values of ρ. This equation is for a very simple unrealistic setting, where all R_i are iid. Given these implausibilities, Table 3.1 shows that these expectations hover around zero, but they may be negative. Thus, $\mathscr{E}(S_P^2)$ under the validity generalization hypothesis appears to be near zero even though it has no simple interpretation. Note importantly that $\mathscr{E}(S_P^2)$ varies both with sample size and ρ, so the expectation is dependent on the unknown but constant ρ. Particularily troublesome, however, is that the expectation of S_P^2 can be smaller, larger, or the same under H_s as under H_g, depending on the unknown ρ. In fact, even negative observed values of S_P^2 can be closer to the expected values of S_P^2 under H_s than under H_g. The following examples make this clear.

EXAMPLE 3.10.1. Let $k = 4$ and $n_i = 10, 40, 50, 180$. Thus, both mean sample size of 70 and the range of n_i are within the ranges often reported (e.g., Carruthers, 1950). Let the corresponding observed $r_i = .4, .4, .3, .2$. From Equation (3.4.8), $s_P^2 = -.0205$. Since the mean of the observed r_i equals .325, if ρ is constant, it seems reasonable to consider $\rho = .325$. Then, under H_g from Equation (3.9.2), $\mathscr{E}(S_P^2) = .00210$. The expectation is different for each different ρ. For example, under H_g with $\rho = 0$, $\mathscr{E}(S_P^2)$ is .00312, while if $\rho = .5$, $\mathscr{E}(S_P^2)$ under H_g is .000952; $\mathscr{E}(S_P^2)$ for $\rho \geqq 0$ is larger for ρ small and smaller for larger ρ, as Table 3.1 shows for the equal sample size case.

However, consider $\mathscr{E}(S_P^2)$ under several H_s. With $\rho_1 = \rho_2 = .45$ and $\rho_3 = \rho_4 = .4$, from Equation (3.9.2) $\mathscr{E}(S_P^2) = .00183$, a smaller expectation than under H_g with $\rho = .325$. Or if $\rho_i = .9, .85, .9, .85, i = 1$ to 4, respectively, then $\mathscr{E}(S_P^2) = .000442$, much smaller than the expected value under any of the H_g hypotheses considered. If $\rho_1 = \rho_2 = .5, \rho_3 = \rho_4 = .4315$, the expected value of

$\mathcal{E}(S_P^2)$ is the same as it is under H_g with $\rho = .325$. Of course, not all choices of ρ_j are equally plausible.

EXAMPLE 3.10.2. Consider the case where all k studies have the same sample size. Let $n = 40$ and $k = 4$, then for each of the following choices of ρ_i the expected value of $\mathcal{E}(S_P^2)$ from Equation (3.9.2) is .0011: (1) $\rho = .13$ (constant); (2) $\rho_i = .88, .85, .85, .8$; (3) $\rho_i = .6, .6, .55, .55$. Other values with the same expectation could be given.

EXAMPLE 3.10.3. The hypothesis testing problem is not nearly so troublesome if H_g is taken as the null hypothesis and H_s the alternative, and such an arrangement is much more congenial to a classical hypothesis testing approach.

For example, consider the following 4 observed $r_i = .7\ .5\ .2\ .5$, with corresponding $n_i = 10, 30, 40, 50$. Then, the observed $s_P^2 = .0216$. The maximum expected value of S_P^2 under H_g occurs at $\rho = 0$, since $\mathcal{V}(R_i)$ is maximum at $\rho = 0$. Thus, for these sample sizes, under H_g with $\rho = 0$, Equation (3.9.2) has value .00360, and it would be even smaller for larger more plausible ρ. Thus, the observed S_P^2 is at least 6 times larger than the largest expected value under any H_g hypothesis, suggesting H_g should be rejected in favor of H_s.

These examples demonstrate that simply observing a smaller near zero value of S_P^2, even if negative, does not allow the unambiguous conclusion to be drawn that there is evidence favoring H_g. The examples also demonstrate that S_P^2 cannot provide the basis for a general composite null test of H_s, of the form "the ρ_i are not all the same" against a composite alternative of the form H_g "all the $\rho_i = \rho$" because the expectation of S_P^2 depends on the unknown ρ, even if ρ is constant for all ρ_i. Thus, to use S_P^2 as the basis for any clear test requires that the ρ_i be specified for both H_s and H_g so they become simple hypotheses. But, this information, even if specified, has limited practical usefulness because the distribution of S_P^2 is unknown.

3.11 The Distribution of S_P^2 and the Power Problem

A major deficiency of the foregoing analyses is that the focus is only on the expected values of S_P^2 under certain possible hypotheses H_g and H_s. Clearly, estimators have variability, and consequently, even if the expectation of S_P^2 were nonnegative, it would be important to know the proportion of the probability mass of the unknown distribution of S_P^2 that can be negative. To do so requires knowledge of the distribution of S_P^2, but that appears analytically unobtainable. Knowing the variability of S_P^2 might help, but analytically $\mathcal{V}(S_P^2)$ is probably beyond reach, even with large sample methods because $\mathcal{V}(S_P^2)$ depends on the eighth moment of R_i and on the unknown covariance between S_*^2 and S_R^2. However, the problem can be studied empirically for moderate sample sizes. Therefore, consider a specific validity generalization testing problem. Let the decision rule be:

1. Reject H_s if $S_P^2 < 0$ and accept H_g.
2. Accept H_s (or fail to reject H_s) if $S_P^2 \geq 0$.

H_g means all ρ_i are the same, and H_s means all ρ_i are not the same.

This rule probably would be considered conservative and perhaps thought to be biased against the validity generalization hypothesis.

EXAMPLE 3.11.1. Suppose the observed $r_i = .3, .4$, and $.5$ are based on $n_i = 30$, 40, and 50, respectively. From Equation (3.4.8), the observed $s_P^2 = -.0094$, and thus H_s is rejected, and H_g accepted. The question is, however, is the decision rule a plausible one?

The distribution of S_P^2 was approximated empirically by simulation for r_i, $i = 1, 2, 3$, with $n_i = 30, 40$, and 50 for selected parameter values of ρ_i.

Figure 3.1 shows smoothed (hand drawn) curves based on histograms of 8100 independent samples of S_P^2, each based on three independent g-distributed correlation coefficients. In Figure 3.1(a), $\rho = .3, .4$, and $.5$, so sampling was under H_s. The empirical distribution has mean .0114, standard deviation .0306, and has range $-.025$ to $.27$. The $\mathscr{E}(S_P^2) = .01094$, obtained

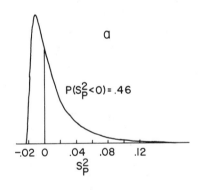

$$P(S_P^2 < 0) = .46$$

a

$$\begin{array}{ccccc} -.02 & 0 & .04 & .08 & .12 \end{array}$$
$$S_P^2$$

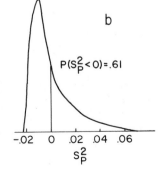

$$P(S_P^2 < 0) = .61$$

b

$$\begin{array}{ccccc} -.02 & 0 & .02 & .04 & .06 \end{array}$$
$$S_P^2$$

FIGURE 3.1. Smoothed histograms of 8100 independent values of S_P^2, Equation (3.4.8); each is based on three independent r_i, each g distributed, with sample sizes $n_i = 30, 40$, and 50. (a) $\rho_i = .3, .4, .5$. (b) ρ constant at .4. The probabilities given are based on empirical frequencies.

from the approximation formulas of Johnson and Kotz (1970b), is very close to the empirical mean. The empirical $P(S_P^2 < 0 \mid H_s) = 3768/8100 = .465$, which is the power or significance level of the test under H_s, the null hypothesis!

Consider the distribution of S_P^2 under H_g, and let the common $\rho = .4$. The corresponding simulation leads to the smoothed curve in Figure 3.1(b). The empirical mean is .00094 with standard deviation of .02. The range is $-.025$ to .22. The approximation $\mathscr{E}(S_P^2) = .00075$. The empirical $P(S_P^2 < 0 \mid H_g) = 4959/8100 = .612$, which is the power under the alternative H_g, which in this case is only moderate.

Of course, the power will change under different choices of parameter values. For example, if $\rho = .2, .4$, and .6, then the empirical $P(S_P^2 < 0 \mid H_s) = .16$, still a very high significance level.

The number of r_i in this example is unrealistically small, and the sample sizes on which each is based are only moderate. But, the example demonstrates that egregious decisions can be made by using what might be thought to be a conservative decision rule given above. Because $P(S_P^2 < 0 \mid H_s)$ can be very large, there can be a very high probability of a Type I error. Thus, to argue that with variance "near zero the hypothesis of situation specifity is rejected" (Schmidt et al., 1979, p. 261) can be very misleading. Note that Figure 3.1 also shows that S_P^2 certainly is no simple variance estimate because roughly half of its probability mass is less than zero. Note also the distribution of S_P^2 depends on ρ_i even if it is constant for all i, a fact noted in Section 3.10.

For another perspective on the power problem see Osburn, Callender, Greener, and Ashworth (1983).

3.12 The Consistency of S_P^2

Another criterion for evaluating an estimator is consistency. An estimator is said to be consistent if it converges in probability (that is approaches) the value of the parameter being estimated as the sample size increases, and it must do so for all possible values of the parameter. Here, sample size refers to k, the number of R_i, while n_i the sample size associated with each R_i is a known parameter.

Consistency is often a minimal defining condition for the suitability of an estimator. It is, for example, Kendall and Stuart's "first criterion for a suitable estimator" (1979, p. 3). Serfling (1980, p. 48) observes that estimators that are not consistent are usually dropped from consideration. If S_P^2 is to estimate a variance quantity, then clearly it is necessary (but not sufficient) that S_P^2 converge in probability to a nonnegative number with increases in k.

The simple (but improbable) example in Section 3.5 shows that $S_P^2 < 0$ for all $k \geq 2$ and independent of n_i. This example demonstrates that S_P^2 cannot be consistent for a very large class of distributions of R. It does not demonstrate that S_P^2 cannot be consistent for a more plausible class of distributions.

Consistency is a large sample property, and as typically defined, it is applicable to situations where the samples are iid. That, of course, is never the practical case in validity generalization because sample size n_i varies over studies, and so the r_i are not identically distributed, as has been noted before. Thus, additional conditions must be imposed on the definition of consistency if it is to apply to the non iid case.

However, it is possible to demonstrate that S_P^2 is not consistent even when the r_i are iid and g distributed.

EXAMPLE 3.12.1. Let R have the same distribution as R_i for all k and let all R_i be independent and g distributed. Thus, the R_i are iid, which fixes $n_i = n$ and $\rho_i = \rho$, and the validity generalization model is Equation (2.3.2).

From Equation (3.7.5),

$$\mathscr{E}(S_R^2) = \mathscr{V}(R) \qquad \text{for all} \qquad k \geq 2, \tag{3.12.1}$$

and from Equation (3.8.2),

$$\mathscr{E}(S_*^2) = \frac{1 + \mathscr{E}(R^4) - 2[\mathscr{E}(R)]^2 - 2\mathscr{V}(R)}{n - 1}. \tag{3.12.2}$$

From these two equations, write

$$\mathscr{E}(S_*^2) \geq \mathscr{E}(S_R^2) = \mathscr{V}(R) \qquad \text{if and only if}$$

$$(n + 1)^{-1}\{1 - 2[\mathscr{E}(R)]^2 + \mathscr{E}(R^4)\} \geq \mathscr{V}(R). \tag{3.12.3}$$

Because the expected values of S_R^2 and S_*^2 have, in this example, expectations independent of sample size k, they will converge in probability by the laws of large numbers to their expectations as k increases (Serfling, 1980, p. 27).

Clearly, the inequality $\mathscr{E}(S_*^2) \geq \mathscr{E}(S_R^2)$ and its equivalent expression in Equation (3.12.3) should never be strict; if they are, then $S_P^2 = S_R^2 - S_*^2$ must converge to a negative value in large samples, and thus could not be consistent.

The moments of R can be computed in the same way as in the previous examples. Table 3.2 shows for selected values of ρ how large n may be before

TABLE 3.2. Values of ρ and maximum values of n for which $\mathscr{E}(S_*^2) \geq \mathscr{E}(S_R^2)$ will hold, Example 3.12.1.

ρ	n	ρ	n
.1	4	.7	15
.2	4	.8	30
.3	4	.85	52
.4	5	.86	61
.5	6	.87	74
.6	9	.89	160

the inequalities in Equation (3.12.3) fail, that is, when the expectation $\mathscr{E}(S_P^2)$ becomes nonnegative. Thus, for example, for $n \leq 15$, the inequality holds for $\rho = .7$; it fails for $n > 15$. Remember that the inequality should be zero for all n since the variability among the ρ_i is zero.

This example shows that S_P^2 is not in general a consistent estimator of variance even in the simplified and unrealistic iid case.

3.13 The Limiting Behavior of S_R^2

Given the interest in simple estimators of the variability among the ρ_i, consider the limiting behavior of S_R^2 as the n_i within each study becomes large but k, the number of studies, remains fixed. Using standard results in the literature (Serfling, 1980), as each $n_i \to \infty$ and as $R_i \overset{\text{P}}{\to} \rho_i$ (read as "R_i converges in probability to ρ_i"), we have

$$S_R^2 \overset{\text{P}}{\to} \frac{\sum\sum_{i<j} (\rho_i - \rho_j)^2}{k(k-1)} = \frac{\sum [\rho_i - k^{-1}\sum \rho_i]^2}{k-1}. \tag{3.13.1}$$

Thus, S_R^2 converges in probability to precisely the quantity that assesses the variability of the ρ_i; if all $\rho_i = \rho$, then S_R^2 converges in probability to zero.

This simple and clear interpretation for S_R^2 occurs because all the variances and covariances vanish as the n_i become large. Equation (3.13.1) could be the interpretation of S_R^2 if each n_i sample size were large. The ordinary sample variance of the R_i then, for large n_i sample sizes unfettered with a substractive correction term, is the desired estimate.

Precisely how large sample size n_i would need to be before Equation (3.13.1) would be a plausible interpretation of S_R^2 can only be answered with Monte Carlo methods. Given the interest in the problem, it might be a simulation worth doing.

Because $S_*^2 \overset{\text{P}}{\to} 0$ as $n_i \to \infty$, for k fixed, $S_P^2 = S_R^2 - S_*^2$ converges in probability to the same value as S_R^2 as the n_i become large. It might be wondered why S_P^2 is not similarly recommended for large sample size n_i settings. One reason is that S_R^2 is always bounded from below by zero, while S_P^2 is not.

3.14 Multifactor Estimation Procedures

The estimator S_P^2 is intended to correct S_R^2 only for error caused by sampling variation in the E_i as discussed in Section 3.4.

The goal of the computer intensive estimation routines is to adjust S_R^2 and S_*^2 for attenuation due to predictor and criterion unreliability and restriction of range as modeled in Equation (2.2.5). Like S_P^2, these ad hoc estimators basically alter functions of differences between S_R^2 and S_*^2 by incorporating functions of the specified finite distributions taken as hypothetical distributions of the attenuating quantities. See Raju and Burke (1983) for distributions that have been used.

Such distributions have sometimes been called prior distributions in an apparent attempt to demonstrate a Bayesian connection. But, there really is no connection with recognized Bayesian procedures. First, the priors have not been incorporated through standard Bayes' rule procedures; they cannot be so incorporated, as noted in Section 3.2. Second, unlike Bayesian procedures in which information in the prior distribution becomes less influential in determining the outcome of the analysis as sample size increases, in validity generalization, the priors may totally dominate the data, as will be seen momentarily.

3.15 A Representative Multifactor Estimator

Burke (1984) summarizes the various estimation equations. He states there are six different estimation procedures, in effect, six estimation equations. All are similar in form, and in general, properties of one estimator are similar to properties of another. Consequently, it is unnecessary to consider each of them.

For illustration, consider the Schmidt, Gast-Rosenberg, and Hunter (1980) estimator. From Burke's (1984) Equation 6, which is slightly different from the original, write

$$S_{\rho'}^2 = \frac{M^2(S_R^2 - S_*^2) - \bar{R}^2 V}{M^4}, \qquad (3.15.1)$$

where $S_{\rho'}^2$ is claimed to be an estimate of $\mathscr{V}(P')$, that is, $\mathscr{V}(P')$ is viewed as the variance of the true validity coefficients, with P' taking on values ρ' as defined in Equation (2.2.6). Thus, one might hope that $\mathscr{E}(S_{\rho'}^2)$ would be approximately $\xi = (k - 1)^{-1} \sum [\rho_i' - (k^{-1} \sum \rho_i)]^2$.

In Equation (3.15.1), M and V are functions of the assumed prior distributions. $M = M_a M_b M_c$, where, for example, "M_a = mean of the square roots of the values of the criterion reliability distribution" (Schmidt et al., 1980, p. 657), and M_b and M_c are similarly defined for prediction reliability and restriction of range distributions. V is the variance of the product of the assumed distributions. Observe critically that M_a, M_b, M_c, and V are fixed known constants. They are not properties of data, although presumably they are intended to reflect such latent data properties; they are thus functionally and stochastically independent of the R_i, and they are bounded. Since, for example, $-1 \le M_a \le 1$, $0 \le M_a^2 \le 1$, and similarly for M_b and M_c and their squares. Consequently, $0 \le M^2, M^4 \le 1$.

Most of the properties of primary concern may be inferred from inspection of Equation (3.15.1). First, the structure of the equation is peculiar. Most estimation equations are functions only of sample statistics; a notable exception, of course, are Bayesian procedures, but Bayesian information does not enter equations in this way. Second, the denominator, M^4, seems very strange. Virtually all sample statistics that are algebraic functions of the observations, such as the sample mean, sample variance, and sample correlation coefficient, have sample size in the denominator. Here, a fixed constant appears. Third, just like S_P^2, it is by no means obvious that $S_{\rho'}^2$ will even be positive.

Now, for any random variable in the interval -1 to 1, its variance must be in the interval 0 to 1 (Note 3.1). Thus, in order for $S_{\rho'}^2$ to be in the variance parameter space, it must also be in that interval. Is it? The answer clearly is that it need not be, and indeed it can take on any value on the real line. To see this, simply allow M or a component of M, for example, M_a, to approach zero. Then, for example,

$$\lim_{M_a \to 0} S_{\rho'}^2 = +\infty \qquad \text{if} \qquad S_R^2 - S_*^2 > 0$$

$$= -\infty \qquad \text{if} \qquad S_R^2 - S_*^2 < 0,$$

and if $S_R^2 - S_*^2 = 0$, the limit may depend on $\bar{R}$.

The point is not that it is plausible to allow M_a to approach zero. The point is, however, that $S_{\rho'}^2$ can become arbitrarily small or large depending on the difference between $S_R^2 - S_*^2$ and the choice of the denominator term. This fact reveals the striking dependency of $S_{\rho'}^2$ on the assumed values of the constant M, which can therefore dominate data. To a large extent, $S_{\rho'}^2$ can be independent of data.

The estimator $S_{\rho'}^2$ can clearly be pathological. Indeed, quite apart from any other features of an estimator that would seem desirable, the reader is challenged to state sensible conditions that would bound $S_{\rho'}^2$ in the interval 0 to 1.

The expected value of $S_{\rho'}^2$ is

$$\mathscr{E}(S_{\rho'}^2) = M^{-2}\mathscr{E}(S_R^2 - S_*^2) - M^{-4}V\mathscr{E}(\bar{R}^2), \tag{3.15.2}$$

where $\mathscr{E}(S_R^2 - S_*^2)$ may be taken from equations in Section 3.9 and

$$\mathscr{E}(\bar{R}^2) = \left[\sum_i \{\mathscr{V}(R_i) + [\mathscr{E}(R_i)]^2\} + 2\sum\sum_{i<j} \mathscr{E}(R_i R_j)\right]\bigg/ k^2. \tag{3.15.3}$$

These expectations have no straightforward interpretation nor do they appear to simplify. But, Equation (3.15.2) makes clear that the expected value is dependent on the chosen fixed constants since these constants factor out in front of the expected values of the random variables, and the expected value changes with different choices of M and V.

Note that $S_{\rho'}^2$ cannot be a consistent estimator of the desired variance even in the iid case. This is because, like expectations, constants factor and thus alter any value to which the random variable $S_{\rho'}^2$ might otherwise converge. Thus, suppose for example X is a random variable and v and θ are constants. Then, if $X \xrightarrow{\text{P}} \theta$, $vX \xrightarrow{\text{P}} v\theta$, and $v + X \xrightarrow{\text{P}} v + \theta$ (cf. Serfling, 1980, p. 19).

Consequently, the value to which $S_{\rho'}$ converges depends on the choices of the fixed constants, and thus $S_{\rho'}^2$ can converge to almost any value desired given the selection of suitable constants.

It is difficult to see how any estimator with the above properties could be regarded as generally useful. An examination of the other estimation equations provided by Burke (1984) will reveal they all possess similar properties.

3.16 A Comment on Z Transformations

The original Schmidt and Hunter (1977) procedure proposed transforming R by Fisher's $Z = .5 \log[(1 + R)/(1 - R)]$ and applying corrections to the variance of Z. This procedure was followed for sometime until certain peculiarities were noted in one validity generalization study (cf. Hunter et al., 1982, p. 42). Subsequently, corrections have been applied to untransformed correlations.

Just how the Z transformation can be conceptually justified remains unclear, particularily in light of recent calls to return to this transformation (e.g., James et al., 1986).

Under the model of Equation (2.3.1), in terms of the latent variables $R = P + E$, the transform becomes

$$Z = .5 \log[(1 + P + E)/(1 - P - E)].$$

How any progress can be made with this transformation is difficult to see. The variance of Z is usually taken as $(n - 3)^{-1}$, but corrections to this variance do not relate to the variances of the underlying latent variables P and E in any clear manner. Furthermore, the variances of P and E are not decomposed because P and E remain locked in the bowels of the transform. Fisher's Z is a marvelous normalizing transform, but it does not appear to be useful for simplifying life in this latent variables context.

4
Summary and Discussion of Validity Generalization

4.1 Summary of Model Properties

A validity generalization random model of the form widely used was defined in Equation (2.3.1) as $R = P + E$; R, P, and E are all random variables; they take on values r, ρ, and e. R is an observed correlation coefficient; P is a latent variable of population correlation coefficients, and E is a latent error variable. P may be viewed as a product of other latent variables that influence testing situations.

Several interesting and critical model properties are elementary consequences of the obvious fact that R is a correlation coefficient and thus is a bounded random variable with $-1 \leq R \leq 1$. This fact imposes critical and heretofore apparently unrecognized constraints on the properties of E and P.

Specifically, (1) the error random variable E must have a different distribution for each fixed $P = \rho$, and E cannot in general be symmetric in distribution about zero. This fact is evident from examination of the joint parallelogram space $\mathscr{P}$ of P and E in Figure 2.1. This space is the space of all possible outcomes of the pairs (ρ, e). (2) Because the joint space of P and E is not a rectangle, Theorem 2.1 states that P and E are dependent random variables. This result does not depend on knowing the joint probability density function of P and E, that is, $f(\rho, e)$, over this space. The result holds as long as $f(\rho, e)$ is positive on the space and zero otherwise.

Although P and E are dependent random variables, they may be uncorrelated depending on how the model is defined. Theorem 2.2 states a sufficient condition for P and E to be uncorrelated. They will be uncorrelated if the model is defined in such a way that for every fixed $P = \rho$, the expectation of E given $P = \rho$ is zero. Whether this property will hold depends on the joint distribution of $f(\rho, e)$ defined over the space. The examples of Section 2.7 are illustrative.

Dependency and correlation are not in general equivalent concepts. It is the dependency of P and E that is a most critical analytical fact and has probably fatal implications for the model because the model becomes, very quickly, analytically intractable.

Virtually all rigorous model development requires that $f(\rho, e)$ be specified, because this distribution is crucial for determining the distribution of R, from which, presumably, it would be possible to develop estimation procedures in a more conventional manner. But, even if $f(\rho, e)$ were specified, the fact that P and E are dependent means that the development of the model would require such complex numerical methodologies it would probably rarely be used. The complexity of the numerical analysis required is substantially elevated simply because P and E are dependent.

A different problem concerns the model's lack of identifiability in the case of the general multiplicative factor model, Equation (2.2.5) or (2.2.7); this issue is considered in Section 4.4.

It is worth noting that several of the analytical results of Chapter 2 made no use of the fact that R is a correlation coefficient. Rather a key elementary fact that forced many of the results is that R is an additive composition [under the model of Equation (2.3.1)] and particularily that R is bounded.

4.2 Validity Generalization and Classical Test Theory

It has been claimed that certain features of the validity generalization model are "exactly comparable to standard formulas in classical measurement theory" (Hunter et al., 1982, p. 43), yet except for the superficial similarity that their latent structures are additive structures each with an error variable, there is little similarity between classical test theory and the validity generalization model, at least as parameterized here.

A crucial fact is again the fact that R is bounded, so P and E are dependent. No dependency is forced in classical test theory because the observed X variable is not a bounded variable. Thus, in classical test theory, T (true scores) and E (error) may be parameterized as independent with $X = T + E$.

Furthermore, in validity generalization, the variable E cannot in general be symmetrical about a mean zero (unless E is degenerate). In addition, the distribution of E must be different for each value of $P = \rho$. In classical test theory the error random variable can typically be assumed to be symmetric about zero and often normal in distribution as well.

In validity generalization, the observed variable R is necessarily a statistic, and in particular a mean; while in classical test theory, X is not necessarily so viewed. Another difference is that R is a biased estimate of ρ, its approximate expectation, while X, when, for example, it is regarded as an ordinary sample mean, is an unbiased estimate of that mean.

4.3 Summary of Estimation Procedures

The validity generalization model is incompletely specified partly because the density of R, $c(r)$, from $R = P + E$ cannot be given. Consequently, it is not possible to proceed with any conventional estimation procedure, either classical or Bayesian, because the joint density function, Equation (3.2.1), cannot be given. Furthermore, the estimation procedures have developed in an ad hoc manner, not in close articulation with any clear model. For this reason, the estimates have no clear relationship with any known model. It is therefore not surprising that the resulting estimates purported to estimate variances have no such clear interpretation and can possess peculiar, sometimes pathological, properties.

In effect, the evolution of the estimation procedures has been just backward of what is desired. Many problems could have been eschewed if a proper probability model had come first, with the estimation procedure following naturally from the model likelihood. This has not happened, so what has resulted is a series of estimation procedures largely detached from a model. The best that can be done is to consider the estimates on their own.

Let R_i, $i = 1$ to k, be correlation coefficients with parameters n_i, sample size, and with population correlation coefficients ρ_i. The central goal of validity generalization estimation is to estimate the variability among the ρ_i, possibly after certain sources of artifactual variation have been eliminated. All estimates, at least all those summarized by Burke (1984), are based on two sample quantities given here in their unweighted forms:

$$S_R^2 = (k - 1)^{-1} \sum (R_i - \bar{R})^2$$

and

$$S_*^2 = k^{-1} \sum (1 - R_i^2)^2 (n_i - 1)^{-1},$$

where $\bar{R}$ is the mean of the R_i. Functions of S_R^2 minus functions of S_*^2 define the general class of estimators in use.

In the simplest case, the estimate S_P^2 is defined by $S_P^2 = S_R^2 - S_*^2$. This estimator has a certain natural intuition associated with it. Namely, it appears to assess the overall variation of the sample correlation coefficients with S_R^2 and subtracts off a component, S_*^2, which purports to represent an average of the estimated sample size variance associated with each R_i. Beyond this intuition, however, it is difficult to find properties of the estimator that are desirable.

By considering the expectation of S_P^2, it is evident it is not simply an estimate of a variance. Sections 3.7 through 3.10 hopefully have made that clear. The estimate S_P^2 is not, in general positive; it can have negative expectation, and even in large samples, under iid sampling considerations, it is not consistent. This means that as k, the number of studies increases, S_P^2 will not converge to the desired variance; it can converge to a negative quantity. Indeed, although the distribution of S_P^2 cannot be given, small-scale simulations indicate that it can have a substantial negative probability mass as Figure 3.1 illustrates.

Validity generalization methods are probably most often viewed from a hypothesis testing perspective. It was this general perspective which Schmidt and Hunter (1977) originally proposed: "If ... [after correction] ... the variance of the distribution of the validity coefficients is essentially zero, the hypothesis [of no variation among the population correlation coefficients or true validities] is confirmed." (1977, p. 530). From a more formal perspective, the two hypotheses might be defined as follows, where H_s denotes a specificity hypothesis, the null hypothesis in this context, and H_g the generalizability alternative hypothesis.

H_s: Among the ρ_i there is at least one inequality.
H_g: All the ρ_i are equal.

If S_P^2 is zero or negative, one decision rule might be to reject H_s and accept H_g, otherwise fail to reject (or accept) H_s.

Even without an unambiguous interpretation, if S_P^2 performed effectively in such a test, then S_P^2 might be redeemed despite its peculiarities. There may be times when it can be useful in this way. However, even with this apparently conservative decision rule above, the significance level of such a test (which will likely always be unknown) can be unreasonably large. Section 3.11 provides an example where the probabilities of wrongly rejecting H_s, when it is true, is about one half. Furthermore, because the distribution of S_P^2 is unknown, it is not possible to assess the power under H_s and H_g a priori. Worse, the distribution of S_P^2 depends on both n_i and the ρ_i, even if (under H_g) the ρ_i are constant.

Suppose, in order to perform an ordinary two-sample t test under the null hypothesis that the population means are equal, it were necessary to specify a point value for the unknown means. If so, the t test would not be very useful. Validity generalization tests face this kind of dilemma.

The complex multifactor model estimators summarized by Burke (1984) are defined by taking functions of the differences of S_R^2 and S_*^2. Collectively they differ from the simple estimator S_P^2 in that fixed, typically data-independent quantities are inserted into the estimation equations. The intended effect of these insertions is to correct for the various statistical artifacts, specifically test and criterion unreliability and restriction of range. These quantities behave as fixed constants in the estimation equations and are not a function of the data. Consequently, the estimators can be dominated by these fixed quantities possibly overwhelming any data. This is not to say there has been any conscious attempt to select such quantities to bias results in any favored direction, only that the properties of the estimators are of this form.

These estimators can have pathological properties. For example, although the variance of any random variable in the interval -1 to 1 must not be larger than 1, the variance estimates can be essentially unbounded in variation. In other ways, they share the same disquieting features of the simple estimator S_P^2 and appear to possess none of the conventionally desirable features of familiar statistical estimators.

4.4 Consistency and Identifiability

The multifactor model estimation schemes (Burke, 1984) attempt to estimate a parameter that, in the general multifactor model, is inherently unidentifiable. Specifically, the attempt is to estimate $\mathscr{V}(P')$, the variance of P', where

$R = P + E$ and, for example, $P = P'A$ in a two-factor model. As noted, these procedures involve inserting data-independent quantities into the estimation equations, quantities that presumably reflect the influence of, in the above example, the random variable A.

It was observed in Section 2.8 that estimation is not meaningful unless the model is identifiable. A model is identified if unique estimates of the parameters are possible; otherwise, it is not identified. It was demonstrated by counterexample in Section 2.8 that unique estimates of $\mathscr{V}(P')$ were not possible in the two-factor model even with observations on P, where $P = P'A$. It was remarked previously that even the simple estimator S_P^2 of Equation (3.4.7) was not consistent in the iid case.

Consistency and identifiability are related concepts, and it is generally recognized that a nonidentified parameter cannot be consistently estimated (e.g., Lehmann, 1983, p. 335; Schmidt, 1983). Because of the significance of this result for validity generalization, the result is stated as a theorem.

Theorem 4.1. *There does not exist a consistent estimator of a nonidentified parameter.* Please see Note 4.1 for the proof.

The result means that $\mathscr{V}(P')$ in the multifactor model with the remaining parameters unknown cannot be consistently estimated. This problem was discussed in Section 2.8. The result does not bode favorably for the future development of estimation under the validity generalization model and casts doubt on the meaning to be associated with use of the estimators in applied settings.

4.5 The Bayesian Connection

In their original paper, Schmidt and Hunter claim to have presented "a Bayesian statistical model" (1977, p. 529), and numerous comments, such as "Bayesian approach," "Bayesian prior," and related expressions claiming a Bayesian affiliation, are sprinkled throughout the 1977 paper as well as subsequent papers.

To say that validity generalization methods are Bayesian is at best misleading. As remarked in Section 3.2.3, the language, but not the concepts of Bayesian analysis, is used. Indeed, by almost any criterion, the procedures are not Bayesian, at least as Bayesian analysis is commonly understood in the statistics literature (e.g., Novick & Jackson, 1974; Berger, 1985).

First, by Lehmann's (1983) criterion, Section 3.2, the procedures are not Bayesian (nor are they classical) because no use is made of the joint density function and, of course, no prior density for the parameters of interest has been parameterized into any model. No use of Bayes' rule appears to have been used in any of the validity generalization methods.

Second, a hallmark of Bayesian methods is that the influence of the prior density for the parameters of interest wears off as more data are obtained. Consequently, Bayesian estimators are generally consistent estimators (cf. Lehmann, 1983, p. 457, Theorem 7.2). This feature does not describe validity generalization priors [i.e., the fixed data-independent quantities in the estimation expressions, e.g., Equation (3.15.1)], the influence of which does not wear off and can dominate data in arbitrary ways.

4.6 Computer Simulation Studies

The primary function of the computer simulation studies has been to explore how well the various multifactor estimation procedures perform in artificial data. Yet the simulation studies that have been done are very unrealistic and are quite peculiar in their design and sampling procedures.

In probably most conventional statistical simulation settings (outside of resampling schemes), a probability model is defined, and samples are obtained under this model or in the case of robust estimation under a near neighborhood model. The desired statistics obtained are compared with the population values, if the population values are known. These features do not describe the simulations that have been done.

Because the validity generalization model has not been fully specified [$f(\rho, e)$ is unknown; cf. Section 2.6], the simulations cannot be said to have taken place under any well-defined validity generalization model. Consequently, just how these simulations relate to any plausible validity generalization model is not clear. Secondly, the studies have ignored sampling error (i.e., variation due to E in model 2.3.7) because "Each of the computer simulation studies only dealt with infinite sample size ..." (Burke, Raju, & Pearlman, 1986, p. 349). Of course, one cannot do studies of "infinite sample size," but the meaning of their statement is clear: sampling error has been ignored. Thus, the data were not real continuously varying correlation coefficients but were discrete values formed by a finite sampling procedure more suitable for other settings.

To illustrate, consider Raju and Burke's (1983) study, which appears typical. Their goal was to compare 5 validity generalization estimation procedures to determine if the point values of the variances of true validities could be recovered. In their Case 1, the true validity coefficient ρ' was fixed; the predictor, criterion, and range restriction artifacts all varied. Each artifact was specified by a hypothetical finite distribution of 100 elements each. For instance, test reliability ranged from .5 to .9 with the elements grouped into 7 bins. The other distributions were similar. As they tell it, "The procedure for producing observed validity coefficients initially involved crossing 100 values of criterion reliability, 100 values of predictor reliability, and 100 values of range restriction effects" (1983, p. 384). Presumably, the square root of each of the products was obtained and then multiplied by the same fixed true validity, ρ'. If so, then in terms of the model here, such products are of the form $\rho'\alpha\beta\delta$ of Equation (2.2.7), where ρ' is fixed for all 10^6 elements.

This procedure is not a simulation at all, because there is no stochastic character to the sample generation scheme; consequently, the estimates are not really estimates, but simply outcomes of numerical calculations. It might be expected therefore that the procedures would return the exact values, but they did not do so. The point values of the ρ' were not in general recovered, and the known zero variances for each fixed ρ' were estimated to be from $-.016$ to $.0148$, depending on the procedure and the fixed ρ' values.

For Case 3, 4 factors were varied. The "observed validity coefficients were produced by randomly selecting without replacement a true validity, criterion reliability, predictor reliability, and range restriction effect from the respective distributions..." (1983, p. 386). Since there were 100 elements in each distribution, there were 100 sampled quantities. Each of these sampled distributions was analyzed with each of the 5 estimation schemes. This finite sampling procedure was replicated 9 times; because there was no variation in sampling error for the R_i, S_*^2 as in Equation (3.15.1) was set to zero.

Even given these unrealistic constraints, the estimates were poor. The average estimated variances of the ρ' were consistently too large and ranged from 1.1. to 1.4 times larger than what they should have been.

Such simulation procedures are considerably at variance with the real world of continuously distributed correlation coefficients, which implies an infinite not finite sampling model, based on sometimes small but certainly varying sample sizes and thus with real sample variation about their parameter values. Consequently, it is difficult to see what real-world generality such studies could have.

The procedures of Raju and Burke (1983) are not unique in the literature; they followed the methods of Callender and Osburn (1980).

5
A Conditional Mixture Model for Correlation Coefficients

5.1 Introduction

Recall again the guiding concern is to characterize histograms of the form of Figure 1.1. A fundamental problem is to assess the variability of the population correlation coefficients underlying such histograms. That appears achievable. In addition, estimates can be given of the number of different ρ that a given data set can support. Once having determined the number of different ρ, maximum likelihood point estimates of each ρ can be obtained.

An estimated variance–covariance matrix associated with the estimates can be obtained. Thus, it is possible to construct confidence intervals for the parameters. Because estimated covariances among the estimates are also available, approximate tests of hypotheses can be constructed.

Through the proposed model, a population model associated with the sample histograms can be specified. Tests of fit of data to the population model, with estimates replacing the model parameters, can be achieved in many cases with ordinary Pearson chi-squared procedures.

This chapter describes the model. Chapter 6 considers point estimation of the model parameters. Applications and examples of proposed solutions are in Chapter 7. Chapter 8 discusses the problem of correcting R for unreliability and range restriction.

Because mixture distributions are central to the model and estimation procedures and because mixture distributions may not be familiar, a brief, relatively informal introduction follows.

5.2 Finite Mixture Distributions

Let $n(\mu, \sigma^2)$ denote a normal distribution with mean μ and variance σ^2. Take a random sample of 4 observations from $n(0, 1)$, that is, a standard normal distribution. Take a second random sample of size 16 from a $n(1, 1)$ distribution, or equivalently, take a second sample of size 16 from a $n(0, 1)$ distribution and add one to each observed value. Now, consider the combined

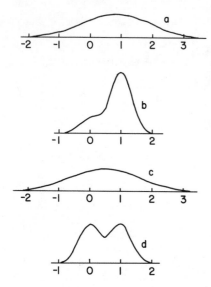

FIGURE 5.1. Four examples of two-component normal mixture distributions. (a) Component means are 0 and 1; component variances are 1 and mixing weights .2 and .8, respectively. (b) The same parameter values as in (a) except both component variances are $\frac{1}{9}$. (c),(d) The parameter values are the same as (a) and (b) except the mixing weights are both $\frac{1}{2}$.

sample of 20 observations. This combined sample has a parent distribution that is a two-component normal mixture distribution. Its density is given by $f(w)$, defined by

$$f(w) = \lambda f_1(w) + (1 - \lambda)f_2(w), \qquad (5.2.1)$$

where $f_1(w)$ is a standard normal density and $f_2(w)$ is a $n(1, 1)$ density. Each component of a mixture is a density (i.e., each component sums or integrates to one), and consequently, associated with each component is a mixing weight. In the example $\lambda = \frac{4}{20}, (1 - \lambda) = \frac{16}{20}$.

Figure 5.1(a) illustrates this normal mixture. While it appears roughly normal, it is in fact quite far from normal. Depending on the choices of component parameter values (both μ and σ^2 in each component may be different) and the associated weights, normal mixture densities can take quite variable shapes. Figure 5.1(b) shows a density with both component variances $\frac{1}{9}$; otherwise, the parameters are the same as in Figure 5.1(a). Figures 5.1(c) and (d) are the same as (a) and (b) except that the mixing weights are equal with $\lambda = \frac{1}{2}$.

The intuition behind Equation (5.2.1) is that with weight λ, observations are taken from f_1, and with weight $1 - \lambda$, observations are taken from f_2. Sometimes the proportions from each component are known and fixed, as in the present example. But, typically the proportions coming from each component are random and unknown and thus λ must be estimated along with the component means and component variances.

While Equation (5.2.1) is a two-component normal mixture, the same basic idea may be extended in several ways. Obviously, there is no need to restrict attention to two components. There may be any number $t \geq 1$ components:

$$f(w) = \sum_{j=1}^{t} \lambda_j f_j(w), \qquad 0 \leq \lambda_j \leq 1, \qquad \sum \lambda_j = 1. \qquad (5.2.2)$$

There is also no need to restrict attention to normal components. The components may be discrete or continuous in distribution, although mixtures of normal distributions are probably the most widely used in applications. There is also no need to restrict f_j to be univariate. If f_j is a bivariate normal density, for example, then f is a mixture of bivariate normal densities, a model which has been useful in other settings (e.g., Thomas, 1983).

Mixtures are among the oldest topics in statistics and normal mixtures were studied by Karl Pearson in 1894. Most recently, interest and research on mixtures has increased as is evidenced by two recent books (Everitt & Hand, 1981; Titterington, Smith, & Makov, 1985). Mixtures arise in various settings and some familar situations may be viewed as mixtures. For example, the familiar t distribution may be viewed as a special kind of normal mixture distribution (Ord, 1972, p. 79), and the marginal distribution of the y (criterion) variable in the standard regression model with normal errors is a mixture of normals distribution.

The validity generalizability–specificity issue with be modeled as a mixture problem. In anticipation of the parameter estimation problem under the proposed model, a tractable replacement for the distribution of R must be found.

5.3 A Modeling Distribution for R

If test and criterion variables are bivariate normally distributed, which will be the assumption maintained throughout, then the distribution of R, $g(r)$, depends only on ρ and n (Patel & Read, 1982, p. 311). Unfortunately, this distribution is not satifactory for modeling purposes because it is complex and cannot be written in closed form. Thus, a replacement for R must be found if it is desired to use R untransformed. Alternatively, R can be transformed to approximate normality using Fisher's Z transformation. Recall that Z is defined by

$$Z = .5 \log[(1 + R)/(1 - R)], \qquad (5.3.1)$$

which is asymptotically normal (e.g., Arnold, 1981, p. 299) with approximate mean

$$\mathscr{E}(Z) \approx .5 \log[(1 + \rho)/(1 - \rho)] = \mu, \qquad (5.3.2)$$

and approximate variance

$$\mathscr{V}(Z) \approx (n - 3)^{-1} = \sigma^2. \qquad (5.3.3)$$

Thus, it would be possible to work with a transformed variable.

There are, however, reasons for desiring to work with the untransformed R. One reason is that R is more familiar and intuitive than a transformed value of R. Therefore, a replacement distribution for R was obtained by defining Z to be exactly $n(\mu, \sigma^2)$, where μ and σ^2 are as defined previously.

From Equation (5.3.1),

$$R = \frac{e^{2Z} - 1}{e^{2Z} + 1},$$

and the distribution of R was obtained by a change of variables (e.g., Hogg & Craig, 1978). That distribution is

$$h(r; \rho, n) = (2\pi\sigma^2)^{-1/2}\{\exp[-(z - \mu)^2/2\sigma^2]\}(1 - r^2)^{-1},$$
$$-1 < r < 1. \qquad (5.3.4)$$

It will also be convenient to denote Equation (5.3.4) both by $h(r)$ and $h(r; \mu, n)$.

This distribution appears to be a very satisfactory distribution for R because $h(r)$ follows the exact distribution of R, $g(r)$, quite closely even for small n and rather extreme ρ, where $g(r)$ is quite skewed. Figure 5.2 shows a worst-case example that graphs $H(r) = \int_{-1}^{r} h(x)\,dx$ and $G(r) = \int_{-1}^{r} g(x)\,dx$ for $n = 15$ and $\rho = .8$. $G(r)$ was obtained from David's (1938) tables while $H(r)$ was obtained by transformation. As can be seen, $H(r)$ is close to $G(r)$. Even for ρ departing further from zero, $H(r)$ is close to $G(r)$ if the sample size is larger. For more moderate ρ, around .4 to .6 or so, the probable case in most employment testing, and with plausible sample sizes, $H(r)$ and $G(r)$ are virtually indistinguishable.

The mean and variance of R, under $h(r)$, obtained by the delta method are approximately

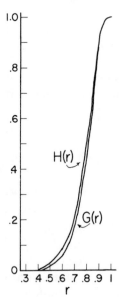

FIGURE 5.2. $G(r)$, the distribution of R, and $H(r)$, the integral of Equation (5.3.4). Parameters are $n = 15$ and $\rho = .8$.

$$\mathscr{E}(R) \approx \mu + \frac{4e^{2\mu}[1 - e^{2\mu}]\sigma^2}{(e^{2\mu} + 1)^3},$$

and

$$\mathscr{V}(R) \approx \frac{\sigma^2 16e^{4\mu}}{(e^{2\mu} + 1)^4}.$$

Thus, $\mathscr{E}(R)$ for practical purposes is about μ, while for ρ in the range of 0 to .8, $\mathscr{V}(R)$ varies correspondingly from about $(n - 3)^{-1}$ to about 13% of $(n - 3)^{-1}$.

Subsequently, when any distribution of R is considered, $h(r)$ will be intended unless otherwise noted. It is well to keep in mind, however, that when, for example, maximum likelihood estimates are considered, they will be maximum under $h(r)$ and not $g(r)$, although in practice they should be quite satisfactory.

5.4 A Mixture Model Distribution for R

For the moment, envision the sample size as fixed at some constant n, and suppose that there is available a collection of different validity coefficients, each arising possibly from a quite different prediction setting. Even with sample size n constant, it is unreasonable to assume, a priori, that each R has arisen from the same distribution $h(r; \rho, n)$. However, whatever the similarities or differences among the distributions $h(r; \rho, n)$ for different R, because the sample size is constant, the parent distributions if they vary can only vary in their ρ values, under the assumption of bivariate normality. Consequently, a natural distribution model to consider should the ρ vary for different R is a mixture distribution defined by

$$h(r \mid n) = \sum_{j=1}^{t} \lambda_j h(r; \rho_j, n), \qquad 0 \leq \lambda_j \leq 1, \qquad \sum \lambda_j = 1. \qquad (5.4.1)$$

Thus, any R with fixed n might have arisen from any one of the t component distributions in Equation (5.4.1). Estimation is considered in the next two chapters. Here, it is simply noted that t, λ_j, and ρ_j are parameters to be estimated.

Equation (5.4.1) is the basic model equation and all later results, for better or worse, are traceable to it. The equation simply reflects our ignorance as to the number of different ρ and, of course, their actual values; Equation (5.4.1) is simply a generalization of the distribution of R if all R had the same parent ρ parameter; if so, then $h(r \mid n) = h(r; \rho, n)$. In other words, Equation (5.4.1) has the spirit: an observed $R = r$ may arise from any one of t possible distributions, each with different ρ_j; the probabilities of it arising from any specific component has probability λ_j.

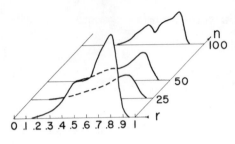

FIGURE 5.3. The conditional mixture model for correlation coefficients with three sample sizes: 25, 50, and 100. At each sample size, there is a three-component mixture distribution with common component parameters $\rho_1 = .5$, $\rho_2 = .7$, and $\rho_3 = .8$ and $\lambda_j = \frac{1}{3}, j = 1, 2, 3$. The functions at each sample size are elements of the joint space and integrate to $\frac{1}{3}$ because it was assumed each sample size is equally probable. The foremost density is the marginal distribution of R, $h^*(r)$, Equation (5.5.1).

Equation (5.4.1) is written as $h(r|n)$ to denote that it is viewed as a conditional distribution, conditioned on n, which so far has been regarded as fixed. Now, regard the sample size as a random variable N with values $n = 2, 3, \ldots, u$. Thus, N is a discrete variable and has mass function

$$f(n) = p_n, \qquad (5.4.2)$$

$0 \leqq p_n \leqq 1$, $\sum p_n = 1$. At each realizable $N = n$, it is assumed that the model parameters t, the number of components, ρ_j, and λ_j are the same. That is the parameters t, ρ_j, and λ_j do not vary with sample size.

The joint density of R and N, $h(r, n)$, is given by

$$h(r, n) = f(n)h(r|n), \qquad -1 < r < 1, \qquad n = 4, 5, \ldots, u. \qquad (5.4.3)$$

Figure 5.3 pictures the complete model joint density in a two-space for the case where there are 3 sample sizes (ignore for the moment, the foremost function in Figure 5.3) $n = 25, 50$, and 100, each with relative frequency

$$p_n = \tfrac{1}{3}, \qquad n = 25, 50, 100$$

$$= 0 \qquad \text{otherwise.}$$

At each of these 3 sample sizes, there are 3 mixture components with identical parameters $\rho_1 = .5$, $\rho_2 = .7$, and $\rho_3 = .8$, and with $\lambda_j = \frac{1}{3}, j = 1, 2, 3$.

The functions shown are elements of this joint space $h(r, n)$. They are not quite the conditional distributions $h(r|25)$, $h(r|50)$ and $h(r|100)$ because they each integrate to $\frac{1}{3}$, but would be precisely the conditional densities with proper normalization, that is, $h(r|n) = h(r, n)/f(n)$, $f(n) = \frac{1}{3}$.

If, in the testing world, all population validities were .5, .7, and .8 with equal weights and all samples were size 25, 50, or 100 in equal proportions, then Figure 5.3 shows the probability model for that world.

Note in Figure 5.3 that each "slice" of the joint density has a different shape even though t and ρ_j are the same. The reason is that the variance for each component of the mixture decreases with increasing n so the individual components for each ρ_j are more prominently in evidence as n increases.

Thus, the model holds that two basic quantities are jointly random: discrete sample size N and continuous R.

5.5 A Parent Distribution for Histograms of R

There are two marginal distributions associated with Figure 5.3. One is the marginal distribution of N, with mass function $f(n) = p_n$. The other is of much more interest, and is the marginal distribution of R. It is obtained by summing Equation (5.4.3) over all n. That is,

$$h^*(r) = \sum_{n=4}^{u} p_n \left[\sum_{j=1}^{t} \lambda_j h(r; \rho_j, n) \right] \qquad (5.5.1)$$

is the marginal distribution of R with density denoted $h^*(r)$. It is thus a t-component, marginal mixture distribution, the result of summing over the sample size variable. The three-component marginal mixture distribution for the joint distribution in the example above is the foremost function in Figure 5.3.

Under the proposed model, Equation (5.5.1) is the parent distribution for histograms, such as Ghiselli (1966) provided, in Figure 1.1.

These marginal distributions for R can be wildly variable in shape. They may be unimodal, multimodal, symmetric, or highly skewed, depending on the parameters of the model. If, for example, all the ρ are relatively similar, then the distribution will typically be unimodal unless the sample size is large. A number of examples appear in Chapter 7.

The main function of the marginal distribution $h^*(r)$, besides providing a clear population model for sample correlation coefficients, is that it provides a model to which data may be fit once parameters of the model have been specified. Unlike most models, however, this parent (marginal) distribution is not the distribution through which estimation takes place. Estimation takes place through the conditional distributions $h(r \mid n)$ or though the joint distribution $h(r, n)$.

5.6 Comment

It is worth noting that none of the model development in Section 5.4 is explicitly dependent on the distribution of R being known. R was regarded as being the outcome of a bivariate normal experiment only in anticipation of the estimation problem, which is made much more complicated if R arises outside a normal theory framework. This is because R, or the components of the distribution of R in the mixture case, is in general no longer only dependent on n and ρ as parameters (Gayen, 1951). Thus, the model development can be largely distribution free. The estimation problem forces the constraints on the component densities of the general mixture model.

6
Parameter Estimation

6.1 Introduction

This chapter provides the basic estimation equations for the conditional mixture model.

Once t, the number of component densities, is specified, λ_j and ρ_j are the only parameters to estimate since $\sigma^2 = 1/(n - 3)$ is always known. The estimates to be given of λ_j and ρ_j are straightforward maximum likelihood (ML) estimates. Examples and applications, the estimation of t, the number of model components, consideration of numerical methods, and construction of confidence intervals are all considered in Chapter 7. The problem of correcting for unreliability and range restriction is considered in Chapter 8.

6.2 Estimation Equations

Let each (R_i, N_i) pair, $i = 1, 2, 3, \ldots, s$, be independent with joint density $h(r, n)$ of Equation (5.4.3). Then, the likelihood function is given by

$$L = \prod_{i=1}^{s} h(r_i, n_i). \tag{6.2.1}$$

If $t = 1$, then $\lambda = 1$ and there is only one μ parameter to estimate. Thus, $\partial \operatorname{Log} L / d\mu = 0$ solves easily for μ and gives

$$\hat{\mu} = \frac{\sum z_i(n_i - 3)}{\sum (n_i - 3)}. \tag{6.2.2}$$

Remembering that $z_i = .5 \log[(1 + r_i)/(1 - r_i)]$, Equation (6.2.2) is the sample size weighted mean. If all n_i are equal, then $\hat{\mu}$ is the ordinary mean.

Because monotone functions of ML estimates are also ML estimates, the ML estimate of ρ is

$$\hat{\rho} = \frac{e^{2\hat{\mu}} - 1}{e^{2\hat{\mu}} + 1}. \tag{6.2.3}$$

In the case where $t \geq 2$, the solution to $\partial \log L/\partial \mu_j = 0$ and $\partial \log L/\partial \lambda_j = 0$, $j = 1,\ldots,t$, do not appear to have closed forms. However, the system can be solved easily using the expectation-maximization (EM) algorithm (e.g., Aitkin & Wilson, 1980).

The solution structure for implementation of this numerical algorithm is conveniently given in the notation of Aitkin and Wilson (1980).

$$\hat{P}(j\,|\,r_i) = \frac{\hat{\lambda}_j h(r_i; \hat{\mu}_j, n_i)}{\sum_j \hat{\lambda}_j h(r_i; \hat{\mu}_j, n_i)} \tag{6.2.4}$$

is the ML estimate of the posterior probability that the ith observation has come from the jth component with sample size n_i. From this basic expression, the ML estimates of μ_j and λ_j, $j = 1,\ldots,t$, are easily given.

The ML of λ_j is

$$\hat{\lambda}_j = \frac{\sum_{i=1}^{s} \hat{P}(j\,|\,r_i)}{s}. \tag{6.2.5}$$

The ML estimate of μ_j is

$$\hat{\mu}_j = \frac{\sum_{i=1}^{s} (z_i/\sigma_i^2) \hat{P}(j\,|\,r_i)}{\sum_{i=1}^{s} \hat{P}(j\,|\,r_i)/\sigma_i^2} \tag{6.2.6}$$

which gives the ML $\hat{\rho}_j$ from Equation (6.2.3).

Finally, the marginal distribution of N will need to be estimated for purposes of fitting solutions to the estimated $h^*(r)$ marginal distribution of R. The ML solution for each $f(n) = p_n$ is

$$\hat{p}_n = \frac{\text{number of samples of size } n}{\text{total number of samples } s}. \tag{6.2.7}$$

7
Examples and Applications

7.1 Introduction

Three assumptions are necessary for meaningful inference under the mixture model. First, the predictor and criterion variables must be assumed to be bivariate normal so that the distribution of R, $g(r)$ may be reasonably approximated by $h(r)$. Second, the individual validity correlation coefficients must be independent. This key assumption is likely to be violated in some data sets as noted earlier (Section 3.3). A third assumption is that for each n_i sample size, there are the same t components in the conditional mixture with the same parameters λ_j and ρ_j.

The solution to the likelihood equations is an application of the EM (expectation-maximization) algorithm. Other numerical approaches such as Newton's procedure could be used. A main virtue of the EM algorithm is that it is an easy algorithm to program. A disadvantage of it is that it can be slow to converge. However, in almost all real data sets encountered so far, convergence speed, as indexed by the number of iterations, has not been a serious problem. Actually, speed of convergence itself can be a useful diagnostic sign. Solutions that converge quickly tend to be satisfactory solutions. In the current implementation all programs were written in APL and run on an ordinary IBM PC. A listing of the main programs is given in the Appendix.

In application, a data vector of r_i, $i = 1,\ldots,s$ is specified along with an identically lengthed vector of associated sample sizes n_i. Starting values of the λ_j and ρ_j (or equivalently the μ_j) are specified in two separate vectors of length t ($\geqq 2$). These starting values are inserted in Equation (6.2.4) and $\hat{P}(j \mid r_i)$ is obtained for each r_i; the $\hat{P}(j \mid r_i)$ are in turn inserted into Equations (6.2.5) and (6.2.6) and new estimates of μ_j and λ_j are obtained. The cycle is repeated, and with each iteration, the likelihood function increases until stationarity or some solution criterion is achieved.

As a solution optimality criterion, rather than increases in the likelihood function L, it is often convenient to use decreases in $-2 \log L$, an equivalent measure. The rationale for using this measure is discussed below.

The estimates obtained on the last iteration are taken as the maximum likelihood (ML) estimates—that is, if the solution is a global and not a local maximum. Because it is never possible to guarantee the solution obtained is in fact the global maximum, new starting values for the parameters must be obtained, and hopefully, the same solution structure will be retrieved. If $-2 \log L$ is found to be smaller with another starting configuration, what was thought to be the ML solution was not the ML solution!

It is important to note that the EM algorithm is supported by convergence theorems that guarantee that the likelihood function will increase or, equivalently, that $-2 \log L$ will decrease (Redner & Walker, 1984).

Typically, increases in the number of t components brings a decrease in $-2 \log L$. Thus, corresponding decreases in $-2 \log L$, possibly penalized by Akaike's criterion, to be described later, may be used as a guide to determine the number of components appropriate for the solution.

7.2 Artificial Data, Example 1

Fifteen correlation coefficients each were obtained from bivariate normal distributions with correlation parameters $\rho_1 = .5$ and $\rho_2 = .7$ and with sample sizes $n = 30$ and $n = 50$. The combined 4 samples represent a sample of size 60 from a two-component mixture distribution at each of 2 sample sizes, with parameters ρ_j and n as just given and $\lambda_1 = \lambda_2 = \frac{1}{2}$. The 60 sample correlation coefficients are given in Table 7.1. Note that the sampling fixed sample size probabilities for each n at one half.

TABLE 7.1. Correlation coefficients for sample data of Example 1.

$n = 50$		$n = 30$	
$\rho = .5$	$\rho = .7$	$\rho = .5$	$\rho = .7$
.49914	.73895	.59636	.51278
.36194	.77191	.41602	.63538
.70059	.71881	.67524	.81969
.43236	.70791	.72719	.74377
.33021	.79210	.54817	.79177
.42927	.63520	.53818	.87931
.53559	.69320	.28862	.69245
.36252	.68159	.65343	.67366
.54311	.62759	.64943	.79513
.50295	.86767	.36162	.75894
.50084	.74420	.37460	.75855
.42675	.73445	.37974	.70064
.37265	.71315	.16635	.37636
.56041	.69910	.18213	.71338
.66003	.77929	.51016	.58921

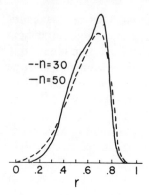

FIGURE 7.1. Two-component mixture distributions for Example 1. The densities are $h(r \mid n = 30)$ and $h(r \mid n = 50)$ with $\lambda_1 = \lambda_2 = \frac{1}{2}$, and $\rho_1 = .5$, $\rho_2 = .7$.

--n=30
—n=50

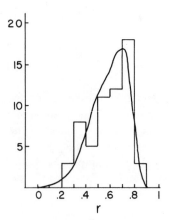

FIGURE 7.2. Marginal mixture density and data histogram for Example 1. Ordinate is histogram frequency.

Figure 7.1 shows the parent two-component mixture distributions, that is, $h(r \mid n)$, Equation (5.4.1), at both sample sizes. Both distributions are unimodal. Should sample size n be sufficiently large, the conditional distributions would be bimodal; a suggestion of developing bimodality is visible for the $n = 50$ distribution.

Figure 7.2 shows both the marginal parent distribution and the realized sample histogram for the $s = 60$ r_i. Sometimes bimodality of a histogram can suggest bimodality of the parent, but in this case the sample bimodality is misleading.

Under a one-component (common ρ) model in which case $\lambda = 1$, the ML estimate of ρ is given by Equations (6.2.2) and (6.2.3); $-2 \log L$ is given by

$$-2 \log L = \sum \text{Log}(2\pi\sigma_i^2) + \sigma_i^{-2} \sum (z_i - \hat{\mu})^2 + 2 \sum \log(1 - r_i^2). \quad (7.2.1)$$

For the specific example at hand, $-2 \log L = -12.72$. Using the model parameter values as starting values for $t = 2$, $-2 \log L = -55.75$, a substantial decrease in $-2 \log L$ when compared with $t = 1$, which suggests $t = 2$ is a far better model for the data. The ML estimates were $\hat{\lambda}_1 = .45$ and

$\hat{\lambda}_2 = (1 - \hat{\lambda}_1) = .55$, where $\lambda_1 = \lambda_2 = \frac{1}{2}$, and the corresponding ρ_j where $\hat{\rho}_1 = .46, \hat{\rho}_2 = .72$, where $\rho_1 = .5$, and $\rho_2 = .7$. These same solution values were obtained from several different starting positions.

A chi-squared goodness of fit test may be used to evaluate the fit of the solution to the data. This measure could provide another index of the suitability of a specified t. Using Cramér's theorem (1946, p. 427), the chi-squared test has, in large samples, approximately a chi-squared distribution with degrees of freedom equal to the number of cell groupings minus one, minus the number of parameters estimated from data. After suitably combining cells so that the expected cell frequencies were about five, under a common ρ model (one parameter estimated), $\chi^2(4) = 104, p < .001$. This result is hardly surprising and signals a very bad fit of solution to the data.

For the two-component solution with $t = 2$, $\chi^2(3) = 3.76, p < .30$, which suggests a reasonable fit of solution to data. There is some uncertainty concerned with determining the degrees of freedom for this test and how the test statistic should be interpreted. This issue is discussed in Section 7.6.1.

Note that the chi-squared value is the fit of the solution to the estimated density $h^*(r)$, where the parameter estimates replace the parameter values. This estimated density is, of course, somewhat different from the population density in Figure 7.2.

This example is fairly typical of other artificial data solutions. Although there were 60 r_i, the sample sizes for the n_i were deliberately not large, and thus the components of the mixture were not sharply distinguished, as Figure 7.1 indicates. The situation makes for a more difficult estimation problem. The procedure seems to work quite well, however, even for a few coefficients if sample size is moderately large. The fact that n_i sample size is always known and largely determines the component variance of $h(r)$ [Equation (5.3.4)] is crucial to the algorithm's apparent success.

7.3 How Many Components in the Mixture?

The problem of determining the suitable number of components in a mixture, particularly when there is no theory for guidance, is important because no other parameters can be estimated until t is specified; thus, the parameter estimates of ρ_j and λ_j must be viewed as conditional estimates, conditioned on the specified number of components t. A variety of procedures for determining the number of components in mixtures have been suggested, graphical and otherwise (cf. Everitt & Hand, 1981), but no one procedure is entirely satisfactory.

One working criterion, a decrease in $-2 \log L$ as t goes from t to $t + 1$, is a fairly common criterion (e.g., Aitkin & Wilson, 1980), and the background motivation for it is useful to note.

Consider the hypothesis $H_0: t = 1$ versus $H_1: t = 2$. If L_0 and L_1 denote the likelihood functions under each of these hypotheses, the likelihood ratio is

L_0/L_1. Then, if Wilk's (1938) theorem could be applied, $-2\log(L_0/L_1)$ is, in large samples, assuming H_0 is true, chi-squared in distribution with degrees of freedom corresponding to the number of parameters by which the model under H_1 exceeds the model under H_0. For example, if $t = 2$, there are 2 additional parameters under H_1 (an additional ρ and associated λ) than under H_0; it might seem that a likelihood ratio chi-squared test could be used to decide the matter.

Unfortunately, Wilk's (1938) theorem does not apply (e.g. Aitkin & Wilson, 1980; Everitt & Hand, 1981, p. 116) because H_1 coincides with H_0 when $\lambda_1 = 1$ then $\lambda_2 = 0$, so the λ_i are on the border of the parameter space. Consequently, a needed regularity condition is not satisfied (namely, there is no derivative on the parameter space border). So, the theorem does not hold. In many practical cases, however, the change in $-2\log L$ can be substantial or very small, so the need for a precise test may not be necessary.

A now commonly used criterion in settings where $-2\log L$ is the basic measure of solution fit is Akaike's Information Criterion (AIC) (Bozdogan, 1987; Sakamoto, Ishiguor, & Kitagawa, 1986). AIC takes decreases in $-2\log L$ as one passes from a simpler to more complex model, but adds a penalty factor for the increase in number of parameters estimated. In the present context, AIC for a t component mixture is defined by

$$\text{AIC}(t) = -2\log L(t) + 2[\# \text{ of free parameters under } t]$$

$$= -2\log L(t) + 2(2t - 1), \tag{7.3.1}$$

where $-2\log L(t)$ denotes $-2\log L$ at t. AIC is an estimate of the expected loglikelihood. The strategy is to select the t component model with the smallest AIC. Although no precise test is possible (Sakamoto, et al., 1986, however, suggest a difference of 1 in AIC is "significant") the emphasis on significance testing may be misplaced anyway. The real concern should be a focus on appropriate model selection.

Another approach to the problem is to use the chi-squared goodness of fit statistic resulting from the fit of the estimated $h^*(r)$ to data. But chi-squared tests are only chi-squared in distribution under the true null model, and obviously not every model can be the true model. Furthermore, there can be substantial change in the observed chi-squared tests depending on how the data are grouped, so this approach is not entirely satisfactory.

7.4 Electrical Workers, Example 2

Dunnette et al. (1982) reported validity coefficients for cognitive tasks and employee ratings for 5 different job classifications in the electric power industry. The r_i were .25, .36, .24, .40, and .32; their corresponding sample sizes were 186, 1385, 460, 258, and 648. These data are but a small portion of a large study.

TABLE 7.2. Estimated posterior probabilities
$\hat{P}(j \mid r_i)$ for electrical workers, Example 2.

i	r_i	n_i	$j = 1$	$j = 2$
1	.24	460	.93	.07
2	.25	186	.61	.39
3	.32	648	.12	.88
4	.36	1385	0	1.0
5	.40	258	.03	.97
		$\hat{\rho}$:	.25	.35
		$\hat{\lambda}$:	.34	.66

If the model assumptions are appropriate, the ML estimate $\hat{\rho} = .34$ under a one-component model with $-2 \log L = -13.34$. For $t = 2$, from a wide variety of starting values, the solution structure yielded $\hat{\lambda}_1 = .34$, $\hat{\lambda}_2 = .66$, $\hat{\rho}_1 = .25$, and $\hat{\rho}_2 = .35$ with $-2 \log L = -15.23$. For example, if the starting values were $\lambda_1 = \lambda_2 = .5$, $\rho_1 = .2$, and $\rho_2 = .4$, convergence to the solution specified was in 5 iterations. There was no further drop in $-2 \log L$ for $t = 3$. The drop in $-2 \log L$ might not seem sufficient to justify a $t = 2$ solution, although AIC(1) $= -11.34$, and AIC(2) $= -9.23$, suggesting that $t = 2$ is the more appropriate solution.

Observe that the r_i .24 and .25 almost coincide with $\hat{\rho}_1 = .25$ while the other 3 r_i might be regarded as clustering around $\hat{\rho}_2 = .35$. It might be thought then that the similarity of the smallest observed r_i to $\hat{\rho}_1 = .25$ suggests the r_i would likely have come from that corresponding $j = 1$ component.

Evidence is provided by considering Equation (6.2.4), which estimates the component membership probability given the data. It can be very useful, once the ML estimates $\hat{\mu}_j$ have been obtained, to insert these estimates back into Equation (6.2.4) and to examine the estimated posterior probabilities.

Table 7.2 reports these probabilities for the five correlations. As can be seen, the most probable component membership agrees with intuition. However, in other examples there can be surprises. $\hat{P}(j \mid r_i)$ depends not just on the similarity of the r_i to the component ρ, but also on the component probability and sample size on which r_i is based. In this example the evidence suggests a two-component model is appropriate for the data.

7.5 Army Jobs, Example 3

This example and the following example are based on US Army classification battery test data from Helm, Gibson, and Brogden (1957), cited and reported in Schmidt and Hunter (1978). There are 2 independent data sets, each based on about 10,500 individuals. The reported correlation coefficients are between a measure of job performance, only noted as "training success measures" (Schmidt & Hunter, 1978, p. 223) and 10 different subtests. The jobs were quite

TABLE 7.3. Iterations of EM algorithm for three-component mixture solution for 35 correlations of US Army jobs, Example 3.

(a) Starting values: $\rho_j = .4, .5, .6$; $\lambda_j = .33, .33, .34$

Iteration	$-2\log L$	$\hat{\lambda}_1$	$\hat{\lambda}_2$	$\hat{\lambda}_3$	$\hat{\rho}_1$	$\hat{\rho}_2$	$\hat{\rho}_3$
1	-56.301	.296	.271	.433	.358	.514	.623
2	-57.663	.288	.306	.404	.347	.517	.629
3	-57.899	.283	.327	.389	.344	.518	.632
4	-57.949	.282	.337	.381	.344	.519	.632
5	-57.962	.282	.341	.377	.344	.519	.633
6	-57.965	.282	.344	.375	.344	.520	.633

(b) Starting values: $\rho_j = .2, .6, .9$; $\lambda_j = .33, .33, .34$

Iteration	$-2\log L$	$\hat{\lambda}_1$	$\hat{\lambda}_2$	$\hat{\lambda}_3$	$\hat{\rho}_1$	$\hat{\rho}_2$	$\hat{\rho}_3$
1	-25.681	.282	.717	0^a	.340	.577	.660
2	-27.040	.308	.692	0	.355	.581	.671
3	-27.197	.318	.682	0	.359	.582	.672
4	-27.222	.322	.678	0	.361	.583	.672
5	-27.227	.323	.677	0	.361	.583	.672
6	-27.228	.323	.676	0	.362	.583	.672

[a] Zero values are approximate; all less than 2×10^{-13}.

variable, ranging from clerks to welders to cooks, and so doubtlessly, the job criteria would be quite variable as well.

This example uses the vocabulary test correlation coefficients for the 35 jobs for the Group A sample; these 35 validity coefficients ranged from .22 to .69, with average .51 and with standard deviation .13. The corresponding sample sizes were uniformly large, ranging from 100 to 767.

For $t = 1$, $-2\log L = 88$, $\hat{\rho} = .60$. For $t = 2$, $-2\log L = -27.23$, a substantial change; for $t = 3$, $-2\log L = -58$, which remained the same for $t = 4$. The decrease in $-2\log L$ as well as with associated values of AIC would suggest a three-component solution is most appropriate.

Table 7.3(a) shows the solution history for one set of starting values that yielded the ML solution. Equally probable λ_j generally represent good initial choices. The three ρ_j starting values were roughly the validity coefficient sample mean and a standard deviation on each side. The solution converges to $-2\log L = -57.97$ in 6 iterations; the ML estimates of the ρ_j and λ_j are the estimates corresponding to iteration 6 and are given in Table 7.3(a).

Table 7.3(b) shows an example of a degenerate solution, which failed to converge to the ML solution, with $-2\log L = -27.23$ in 6 iterations. It illustrates what may happen with implausible starting values. In this case, a $\rho_3 = .9$ seems quite implausible, because it exceeds the largest observed r_i by .21. The result is that the solution converged to $-2\log L = -27.23$, the $t = 2$ solution.

Figure 7.3 shows a histogram of the 35 validity coefficients and the estimated marginal parent density $h^*(r)$ from Equation (5.5.1), where the

FIGURE 7.3. Sample histogram and estimated marginal density for data of Example 3. Ordinate is histogram frequency.

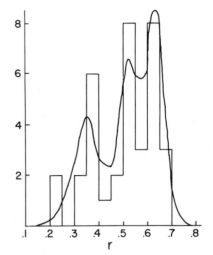

estimates of ρ_j, λ_j, and p_n replace the parameter values. In this case, $\hat{p}_n = \frac{1}{35}$ for each different sample size n because each sample size is different. The fitted density appears to follow the histogram fairly well. A chi-squared fit is difficult to justify in this case, however, because after combining cell frequencies to obtain expected values near 5, there were insufficient degrees of freedom left for fitting.

While there seems little reason for wishing to do so, because point estimates of the ρ_j are available, if an estimate of the variability of the ρ_j is desirable, it is available from

$$\hat{V}(\rho) = t^{-1} \sum \left(\hat{\rho}_i - \sum \frac{\hat{\rho}_i}{t} \right)^2. \tag{7.5.1}$$

This estimate is the conditional mixture model's analog of S_ρ^2 for validity generalization, Equation (3.4.7). The estimate $\hat{V}(\rho)$ is doubtlessly not an ML estimate, even under $h(r, n)$, but it is a function of estimates that under $h(r, n)$ are ML estimates. Because ML estimates are known to be consistent estimates (Norden, 1972), $\hat{\rho}_j$ converges in probability to ρ_j as the number of correlation coefficients, s, becomes large (see also Section 8.5). Consequently, $\hat{V}(\rho)$ converges to $t^{-1} \sum (\rho_i - \sum \rho_i/t)^2$ under $h(r, n)$. This does not imply it will be consistent under the true density, but certainly it should be a good estimate. $\hat{V}(\rho) = .119$ for this example, with $t = 3$.

7.6 Army Jobs, Example 4

This example uses the same Group A US Army vocabulary data as in Example 3, combined with the presumably independent but otherwise identical data set, Group B (Schmidt & Hunter, 1978). Thus, the 70 validity coefficients have mean .51, standard deviation .14, and range from .11 to .75. Sample sizes ranged from 100 to 767.

For $t = 1, 2$, and 3, $-2 \log L$ was 273.64, -25.07, and -96.53; there is no further change in $-2 \log L$ for $t = 4$. Convergence for $t = 3$, the accepted solution, was somewhat slower than for Example 3, but $-2 \log L$ seemed to stabilize within 6 to 15 iterations, depending on the starting values. When the decrease in $-2 \log L$ from a preceding iteration was less than .001, the routine was terminated and the solution accepted. With this decision rule, the parameter estimates typically remained stable to 3 places and usually more.

Equal probabilities for the λ_j with ρ_j values based on descriptive statistics of sample data, inspection of the sample histogram, or setting the ρ_j values at about equally spaced intervals within the range of observed r_i have been found to be safe starting configurations for an initial solution. There is, however, no substitute for experimenting with the different starting values and observing both the decrease and rate of decrease of $-2 \log L$ and noting corresponding changes in the estimated parameters.

The estimates for $t = 3$ were $\hat{\lambda}_j = .279, .291$, and .430 and $\hat{\rho}_j = .337, .518$, and .639 for $j = 1, 2$, and 3, respectively. The $\hat{\rho}_j$ are very close to the values obtained in Example 3, shown in Table 7.3, and the weights show reasonable similarity as well. Thus, the solution structure appears to replicate the results of Example 3. An independent analysis of the Group B data alone, just as in Example 3, revealed a very similar solution as well.

7.6.1 A FITTING PROCEDURE

Figure 7.4 parallels Figure 7.3 and shows the estimated parent marginal density $h^*(r)$ and the data histogram. With the data grouped into 12 cells, the chi-squared fit of the solution to data was $\chi^2(6) = 25.72$, $p < .01$, and thus the fit of the solution to the data does not appear to be very good. However, it is important to consider this result.

Examining the individual cells revealed that the first two cells and the last cell contributed over half of the total chi-squared contribution. So, the fit appears poor in the tails. Another consideration is that in practice it can be difficult to group the data in such a way that there will be a sufficient number of cells to allow a test to be performed and at the same time keeping cell expected frequencies around five or so. For the frequency histogram in Figure 7.4, there are still cells with low expected frequency, even below one. So, the probabilities assigned to the observed chi-squared values must be viewed with caution whether they are large or small.

When determining the degrees of freedom for the test statistic, only the estimates of $\hat{\lambda}_j$ and $\hat{\rho}_j$ were counted as estimated parameters. Thus, when $t = 3$, 5 parameters were estimated (one $\hat{\lambda}_j$ is obtained by subtraction), so again applying Cramér's theorem (1946, p. 427), there are 6 degrees of freedom for fitting.

In fact, however, to fit the empirical histogram to the estimated $h^*(r)$, it is necessary to estimate the marginal frequencies p_n from Equation (6.2.7). In this

FIGURE 7.4. Sample histogram and estimated marginal density for Example 4. Ordinate is histogram frequency.

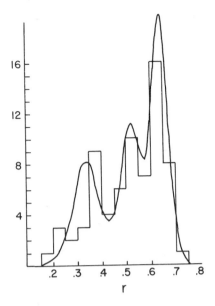

case, there were 59 different sample sizes from among 70 possible; thus, 58 more parameters were estimated! One quickly runs out of degrees of freedom to perform a test! It seems essential therefore to view the fitted chi-squared test as a conditional test, conditioned on (1) the observed marginal sample size distribution and (2) the number of components, t.

Furthermore, variation in $\hat{p}_n$ clearly alters the shape of the estimated $h^*(r)$ and, consequently, the fit of the solution to data. Therefore, a large chi-squared value does not necessarily mean that the number of components accepted is inappropriate or that the model estimates of major interest, $\hat{\rho}_j$ and $\hat{\lambda}_j$, are poor ones. The bottom line is that it may be difficult to discern what may be in error should the chi-squared fitted value be large.

The happy note is that these problems pertain only to the fitting of the solution to the data; they are not directly relevant to the solution estimates of main concern, the $\hat{\rho}_j$ or $\hat{\lambda}_j$, which do not depend on the estimates $\hat{p}_n$.

7.6.2 CONFIDENCE INTERVALS

There are two possible strategies for constructing confidence intervals for the parameters λ_j and ρ_j. One approach is to obtain asymptotic variances of the estimates. A second approach is to utilize resampling schemes, in particular the bootstrap (Efron & Tibshirani, 1986). The bootstap is a considerably more flexible tool for the problem at hand, as will be seen. Its primary disadvantage is that it is computationally intensive. It is, however, trivial to implement.

The bootstrap procedure is as follows. (1) Let (r_i, n_i), $i = 1, \ldots, s$, be a vector of observed coefficients and their sample sizes. (2) Draw a sample of size s from the vector with replacement. This is the bootstrap sample. (3) Compute the parameter estimates for the desired fixed t component mixture in the usual way, obtaining a vector of bootstrap estimates $a_1^* = (\hat{\lambda}_1^*, \ldots, \hat{\lambda}_t^*)$, $c_1^* = (\hat{\rho}_1^*, \ldots, \hat{\rho}_t^*)$, where $\hat{\lambda}_j^*$ and $\hat{\rho}_j^*$ are the estimates based on the bootstrap sample. (4) Repeat (2) and (3) B times obtaining $a_1^*, \ldots, a_B^*$, and $c_1^*, \ldots, c_B^*$. According to Efron and Tibshirani (1986), B in the range of 50 to 200 is adequate for most cases. However, the setting at hand is somewhat different from the typical bootstrap as will be noted momentarily, and thus probably the larger B is the better. (5) From the a^* and c^* vectors, compute the ordinary sample variance–covariance matrix. For example,

$$\frac{\sum_{b=1}^{B} (\hat{\rho}_{1b}^* - \bar{\rho}_1^*)(\hat{\rho}_{3b}^* - \bar{\rho}_3^*)}{B}, \tag{7.6.1}$$

where $\bar{\rho}_1^*$ is the mean of the bootstrap values $\hat{\rho}_{1b}^*$; Equation (7.6.1) is the bootstrap sample covariance estimate of the estimates of ρ_1 and ρ_3.

Once having obtained ML estimates $\hat{\lambda}_j$ and $\hat{\rho}_j$, these values are inserted as starting values for each bootstrap replication. The solution criterion for the bootstrap estimates is to iterate until the decrease in $-2 \log L$ is less than some critical value, such as .001. Thus, the number of iterations for each bootstrap replication is random. An alternative criterion, which typically takes less computer time but which seems to work well in practice although it may be conceptually less defensible, is to fix the number of iterations for each replication. This number can be determined from prior considerations, such as original convergence speed.

For the example at hand, the bootstrap estimated variance–covariance matrices for the estimates $\hat{\lambda}_j$ and $\hat{\rho}_j$ based on $B = 150$ are given in Tables 7.4 and 7.5. Let it be observed that these estimates are not the sort of thing to be computed over a coffee break. A microcomputer initialized in the evening may be expected to return results sometime the following afternoon! Even mainframe time can be nontrivial when convergence criteria are integral parts of the bootstrap procedure.

In conventional applications, the bootstrap is a tool for iid random variables and in a wide range of such applications it works well (Efron & Tibshirani, 1986). If the (r_i, n_i) pairs are regarded as independent samples from

TABLE 7.4. 1000 times the estimated variance–covariance matrix of $\hat{\lambda}_j$, based on 150 bootstrap replications, Example 4.

	$\hat{\lambda}_1$	$\hat{\lambda}_2$	$\hat{\lambda}_3$
$\hat{\lambda}_1$	2.971	-1.469	-1.501
$\hat{\lambda}_2$	-1.469	3.954	-2.485
$\hat{\lambda}_3$	-1.501	-2.485	3.986
$\hat{\lambda}_j$	.279	.291	.430

TABLE 7.5. 10,000 times the estimated
variance–covariance matrix of $\hat{\rho}_j$, based on
150 bootstrap replications, Example 4.

	$\hat{\rho}_1$	$\hat{\rho}_2$	$\hat{\rho}_3$
$\hat{\rho}_1$	4.072	.7409	−.09294
$\hat{\rho}_2$	.7409	1.404	.3234
$\hat{\rho}_3$	−.09294	.3234	.6871
$\hat{\rho}_j$	.337	.518	.639

$h(r, n)$ of Equation (5.4.3), then the pairs would be regarded as iid from $h(r, n)$, and thus the application of the bootstrap here puts the problem much closer to the traditional applications of the bootstrap. It is worth noting, however, that the bootstrap seems to work well even in cases for which there is no rigorous conceptual justification (Efron & Tibshirani, 1986).

Consider Tables 7.4 and 7.5 and observe that the variances associated with the $\hat{\rho}_j$ are smaller than for the $\hat{\lambda}_j$, which is typical. Also note that the larger the $\hat{\rho}_j$, the smaller the variance, an expected result.

An approximate 95% confidence interval (CI) may be constructed in the usual way: take the parameter point estimate and add and subtract two bootstrap standard errors. The standard errors are simply the square roots of the diagonal elements of Tables 7.4 and 7.5. The point value estimates and the associated confidence intervals for λ_j and ρ_j are given in Table 7.6.

Confidence intervals may also be constructed from the histogram of bootstrap estimates for each parameter. Select the 2.5% and 97.5% (or closest) values of the histogram to construct a 95% CI. The 4th and 147th largest bootstrap values for each estimated parameter appear in the right portion of Table 7.6. In this example, the two procedures provide very similar intervals.

Because a true 95% probability interval would most likely be asymmetric, the histogram potentially could provide more accurate intervals, but at the cost of a larger bootstrap sample. B is often 1000 for histogram estimates

TABLE 7.6. Bootstrap 95% CIs for λ_j and ρ_j, Example 4.

j	$\hat{\lambda}_j$	Standard error intervals	Histogram intervals
1	.279	.170–.388	.176–.410
2	.291	.165–.417	.140–.416
3	.430	.304–.556	.284–.536

j	$\hat{\rho}_j$	Standard error intervals	Histogram intervals
1	.337	.297–.377	.288–.376
2	.518	.494–.542	.493–.541
3	.639	.622–.656	.622–.655

(Efron & Tibshirani, 1986). $B = 150$ is probably too small to be of general usefulness.

7.6.3 APPROXIMATE TESTS

Given that the estimated covariances are available in Tables 7.4 and 7.5, it is tempting to propose hypothesis testing procedures in a natural but not necessarily rigorous way. With this proviso and with the recognition that the probability levels must be regarded as only roughly approximate, the following procedure is straightforward.

To test the null hypothesis that $\rho_2 = \rho_3$ against the alternative $\rho_2 \neq \rho_3$, compute $\hat{\rho}_2 - \hat{\rho}_3$ and divide the result by the square root of the sum of the bootstrap variances for each estimate minus 2 times the associated bootstrap covariance. Thus, $(.518-.639)/.012 = -10.08$, which under the null hypothesis could be expected to be referenced, at least approximately, to a standard normal table. If so, then clearly the hypothesis that $\rho_2 = \rho_3$ is rejected and similarily so for the other 2 pairs. The same procedure can be used for the $\hat{\lambda}_j$, but none of those pairwise differences appears significant.

7.6.4 POSTERIOR PROBABILITIES $\hat{P}(j \mid r)$

Example 2 (Section 7.4) provided an illustration of one use of the estimates of the posterior probabilities $\hat{P}(j \mid r_i)$, the probability of component j membership given an observed r_i, or the probability that a particular r_i has been sampled from component j. These probabilities may be useful in a variety of other ways and might suggest avenues of inquiry that would not otherwise be considered.

In the present example, there are 35 different Army jobs and for each there are 2 independent r_i, each based on a rather large sample. It might be expected that the 2 sample r_i for the same job would reveal roughly the same posterior probability of membership. In general, this is the case, but there are some peculiarities. Table 7.7 shows the $\hat{P}(j \mid r_i)$ for selected jobs. These estimates were obtained from Equation (6.2.4), with the ML estimates inserted as the estimated values.

First, consider the field radio repairmen. Both r_i coefficients are moderate and somewhat disparate. Table 7.7 reveals these two correlations clearly appear to have different parent ρ_j, and in both cases with very high probability. This observation suggests that perhaps there may be some systematic differences between the two populations represented by the samples. The same hypothesis might also be true for dental assistants as well. Finally, consider the last two jobs in Table 7.7. For one job, the two r_i are in the forties, for the other the two, r_i are in the fifties. Is there some systematic difference between these two jobs? The posterior probabilities suggest perhaps not; all four samples most likely have arisen from a population with the same common ρ, about .52. More specifically, jobs with similar component memberships might well be candidates for validity generalization.

TABLE 7.7. Estimated probabilities $P(j \mid r_i)$ that the ith correlation came from the jth component for selected Army jobs, samples A and B, Example 4.

Job	N	Group	r_i	$j = 1$	$j = 2$	$j = 3$
Field radio repairman	280	A	.54	0	.94	.06
	481	B	.41	.94	.06	0
Dental assistant	367	A	.51	0	1.0	0
	200	B	.65	0	.01	.99
Criminal investigator	192	A	.62	0	.08	.92
	106	B	.75	0	0	1.0
Armour track vehicle maintenance	248	A	.48	.03	.96	0
	248	B	.45	.22	.78	0
Fuel and electrical systems repairman	523	A	.51	0	1.0	0
	522	B	.55	0	.98	.02
			$\hat{\lambda}$:	.28	.29	.43
			$\hat{\rho}$:	.34	.52	.64

Note: probabilities are rounded.

Such procedures obviously deserve to be used with caution; clearly, the differences might simply represent sampling differences. Incidentally, if it were desired, confidence intervals could be constructed about these estimated probabilities using the same bootstrap replications that were used to construct confidence intervals and tests for the main model parameter estimates.

7.7 College Grades, Example 5

Novick and Jackson (1974, p. 21) reported correlation coefficients between the American College Testing Program's ACT English subtest and first-year college grades for 22 community colleges. The sample sizes were modest, ranging from 26 to 184, and averaged 61. The coefficients ranged from .14 to .70 with mean .42 and standard deviation .157.

Analysis revealed the data could well have arisen from a distribution with one common component; $-2 \log L = -24.16$ under a one-component model. Its values for $t = 2$ and $t = 3$ remained almost the same, never dropping below -25. Consequently, a one-component solution was accepted.

The estimate $\hat{\rho} = .446$ follows from Equations (6.2.2) and (6.2.3). Following 201 bootstrap replications, the standard error of the estimate was .013, so an approximate 95% CI for ρ is .420 to .472. The data histogram and estimated marginal parent density under the model are shown in Figure 7.5. The fit of the solution to the data appeared satisfactory with $\chi^2(1) = .79$, which is not significant.

Another approach to the problem of testing for homogeneity of the ρ_j from the r_i is given by Kraemer (1975). Unfortunately, her procedure has no closed form solution, so an iterative solution is necessary. However, she provides asymptotically chi-squared tests under the homogeneity hypothesis.

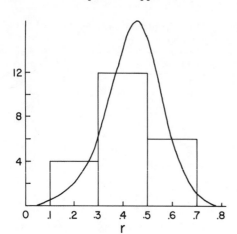

FIGURE 7.5. Sample histogram and estimated marginal density for Example 5. Ordinate is histogram frequency.

7.8 Law School Test Scores and Grades, Example 6

A large data set summarizing the usefulness of several variables on predicting law school grades was presented by Schrader (1977). Data at 4 time intervals were reported for 150 law schools. The criterion variable was the student's first-year law school grade average. Four predictor variables were reported, but only the Law School Admission Test (LSAT) coefficients are used here.

The data across schools are not directly comparable, and for data within schools there are dependencies. For example, it was common to report coefficients for the 1973 entering class as well as the coefficient for the combined 1973 and 1974 classes. Clearly, there is a part–whole problem here. Thus, the entire data set is difficult to characterize. Nonetheless, Linn, Harnisch, and Dunbar (1981) reported a validity generalization analysis based on the entire 726 LSAT score and grade correlation coefficients reported.

For the present analysis, 150 independent validity coefficients were selected, one for each of the schools. A decision rule dictated which coefficient to select (see Note 7.1), but in general, the most recent results reported were selected, and for the majority of schools this meant the combined years of 1973 and 1974, the most recent years reported.

The sample sizes were typically large, ranging from 75 to 1117, with average size 352. The observed coefficients ranged from .04 to .62 with mean .344 and standard deviation .120.

Solutions were obtained for the $t = 1$ common component solution to $t = 5$. Table 7.8 reports the parameter estimates, $-2 \log L$ for $t = 1$ to $t = 4$, along with AIC from Equation (7.3.1). When the change in $-2 \log L$ was less than .001 from one iteration to the next, the routine terminated and the solution was accepted. When $t = 5$, $-2 \log L$ was about -210, the approximate value for $t = 4$, suggesting there is no basis for a five-component solution.

TABLE 7.8. Solution criteria and parameter estimates for models with components 1 to 4, Example 6.

Components	$-2\log L$	AIC	Estimates $\hat{\lambda}_j, \hat{\rho}_j, j = 1\text{--}4$
$t = 1$	338.6	-340.6	$\hat{\rho} = .367$
$t = 2$	-124.8	-118.8	$\hat{\lambda}_j$: .583, .417
			$\hat{\rho}_j$: .269, .462
$t = 3$	-197.3	-187.3	$\hat{\lambda}_j$: .147, .518, .335
			$\hat{\rho}_j$: .161, .315, .476
$t = 4$	-209.9	-195.9	$\hat{\lambda}_j$: .101, .332, .308, .258
			ρ_j: .142, .276, .368, .490

In general, solution convergence was rapid. There was invariably a large precipitous drop in $-2\log L$ from almost any starting position for any $t \geqq 2$ compared with the one-component $-2\log L$.

The change in $-2\log L$ between $t = 3$ and $t = 4$ would, if asymptotic chi-squared tests were appropriate, be significant with two degrees of freedom. AIC for $t = 4$ is the smallest, suggesting $t = 4$ is the best solution. Bootstrap variance–covariance estimates have not been provided simply because of the extensive computations required.

The estimated marginal distribution for $h^*(r)$ for the four-component solution and the sample histogram based on the 150 coefficients are given in Figure 7.6. The fitted curve follows the histogram reasonably well, $\chi^2(4) = 8.9$, $p < .10$. As with Example 3, the cells associated with the tails of the distribution tended to contribute more heavily to the total chi-squared contribution. Given the potential power of the test, 150 data points based on a combined sample size of 52,783, the fit does not appear to be a bad one.

It has been claimed that "there is strong evidence of validity generalization for the LSAT" (Linn et al., 1981, p. 287). The present analysis might well

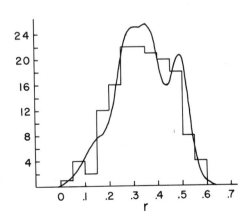

FIGURE 7.6. Sample histogram and estimated marginal density of four-component solution of Example 6. Ordinate is histogram frequency.

suggest otherwise, even without considering the influence of unreliability and range restriction.

Table 7.8 shows that for the $t = 4$ solution 10% ($\hat{\lambda}_1 = .101$) of the schools had estimated ρ_1 of about .14, while for about 25% of the schools, the estimated ρ_4 was around .50. Such differences do not appear to be trivial between-school differences. Furthermore, such schools may be identified by considering the $\hat{P}(j | r_i)$ if desired.

8
Artifact Corrections and Model Assumptions

8.1 Artifact Corrections

As long as the data satisfy the model assumptions, the estimates should be satisfactory and plausible. No restrictions need to be placed on how the model parameters ρ_j and their estimates are viewed. They may be regarded as a product of other parameters if desired. However, in many practical testing settings, measurement unreliability and range restriction are important considerations and cannot be regarded as negligible. The obvious question arises as to how the model or procedures can be altered to allow for corrections of these influences. There are several possible approaches to the problem. Because corrections for unreliability and corrections for range restriction present different kinds of problems, each will be considered separately.

8.1.1 CORRECTING R FOR UNRELIABILITY

The obvious first thought for correcting unreliability of predictor and criterion measurement is simply to use the familiar correction for the attenuation formula, which replaces the parameter values with suitable estimates or guesses in Equation (2.2.4). Following Equation (2.2.6), such corrections of R take the form $R/(\alpha\beta)$, $0 < \alpha\beta \leq 1$. Once the corrections have been made, then estimation would proceed as before.

Although widely used and long viewed as the practical approach to the problem of correcting for unreliability, the procedure is at best very crude and can lead to substantial error even if α and β are known exactly. The problem is that the correction is being applied to random quantities, not parameter values, and the distribution of R is altered by such corrections.

Suppose R has distribution $h(r)$ or $h(r \mid n)$ when $t = 1$, and the random variable $R_a = R/(\alpha\beta)$. Then, $-1 < R < 1$, and $-1/(\alpha\beta) < R_a < 1/(\alpha\beta)$. Assuming α and β are fixed, R_a has density

$$h_1(r_a) = \alpha\beta[1 - (\alpha\beta r_a)^2]^{-1}(2\pi\sigma^2)^{-1/2}\exp\{[-z(\alpha\beta r_a) - z(\rho)]^2/2\sigma^2\}, \quad (8.1.1)$$

where $z(w) = .5\log[(1 + w)/(1 - w)]$ and $\sigma^2 = (n - 3)^{-1}$. Division of R by the

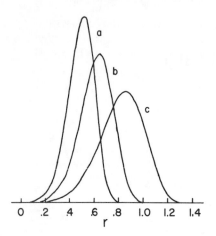

FIGURE 8.1. The distribution of R and R_a, $\rho = .5$, $n = 50$. (a) $h(r)$. (b) $h_1(r_a)$ with $\alpha\beta = .8$. (c) $h_1(r_a)$ with $\alpha\beta = .6$.

correction $\alpha\beta \neq 1$ puts R_a on an interval other than -1 to 1. Figure 8.1 shows the distribution of R ($\alpha\beta = 1$) and R_a for $\alpha\beta = .8$ and .6 when $\rho = .5$ and $n = 50$. The figure reveals that the distribution of R_a can be quite different from the distribution of R, $h(r)$. For instance, if the correction $\alpha\beta$ departs sharply from one, the distribution of R_a can have a large probability mass above one.

The consequences of such corrections are predictable. If the $\alpha\beta$ correction to each R is severe, so that the distribution of $R/(\alpha\beta)$ is strikingly different from the distribution of R, then estimates constructed under the distribution of R, $h(r)$, could be very poor and possibly grossly misleading.

Alternatively, one might correct R by $R/(\alpha\beta)$ assuming $h_1(r_a)$ is the basic model density and proceed with estimation as before but under $h_1(r_a)$. However, this suggestion is not a good one because Equation (8.1.1) does not model reality well. After all, whether test and criterion reliabilities are known or not, and regardless of their values, correlation coefficients are always bounded in -1 to 1, so it would be strange working with a model that can take on values outside this interval. Furthermore, under conventional normality assumptions, R will always be g distributed regardless of reliability, so $h(r)$ should be a suitable model density. The problem with conventional attenuation correction procedures is that they are not introduced in a proper way.

The reliability problem may be viewed in the following way. The random variable R, on which observations are made, has parent correlation ρ and density $h(r; \rho, n)$; ρ is attenuated with unreliability. It is desired to correct R so that modeling takes place under a density $h(r; \rho', n)$, where, from Equation (2.2.6), $\rho' = \rho/(\alpha\beta)$. Thus, the problem is to transform R, which has density $h(r; \rho, n)$ before correction, to a random variable that has density $h(r; \rho', n)$ after correction. The following illustrates the general solution to this problem.

Let X and W be any two continuous random variables with (cumulative) distribution functions $F(x)$ and $G(w)$ and with G^{-1} the inverse of G. It is desired to transform X so that it has the distribution of W. To do so, transform X by $G^{-1}(F(X))$; the random variable $G^{-1}(F(X))$ has the distribution of W. Transforming X by its distribution function results in a distribution that is uniform on zero to one (e.g., Hogg & Craig, 1978, p. 126), while G^{-1} transforms the uniform variable to a variable with the distribution of W. A proof of this result is given in Thomas (1982).

To apply this result, each observed r is first transformed by its distribution function $H(r; \rho, n) = \int_{-1}^{r} h(t; \rho, n)\, dt$, and is then transformed back to a new corrected correlation coefficient by $H^{-1}(r; \rho', n)$. In practice, each observed r would be corrected in this way.

There are practical problems to consider with this approach because the parameters ρ and ρ' would need to be specified. One possibility is to first obtain estimates $\hat{\rho}_j$ without correcting for unreliability. Then, given knowledge of $\alpha\beta$, obtain new estimates $(\alpha\beta)^{-1}\hat{\rho}_j = \hat{\rho}_j'$. These estimates, $\hat{\rho}_j$ and $\hat{\rho}_j'$, when suitably matched with each observed validity coefficient, possibly using the method outlined in Section 7.6.4, could become the values to insert in the transformation equations. New corrected correlation coefficients would be obtained, and the parameters of the mixture reestimated.

Because neither H nor its inverse H^{-1} can be written in closed form, the procedure is necessarily numerical, but this difficulty presents no fundamental problem.

8.1.2 CORRECTING ESTIMATES $\hat{\rho}$

If it were assumed that for all R_i the unreliability attenuations were at least roughly the same, then simply correcting the estimates $\hat{\rho}_j$ by $\hat{\rho}_j/(\alpha\beta)$ might be a useful approximate approach. Note that because a length based on two times the bootstrap standard error, Section 7.6.2, constructs about a 95% CI for ρ_j, then $2/(\alpha\beta)$ times the bootstrap standard error of $\hat{\rho}_j$ when added to and subtracted from $\hat{\rho}_j'$ will construct about a 95% CI for ρ_j'.

8.1.3 CORRECTING FOR RESTRICTION OF RANGE

The expression restriction of range is probably unique to psychometrics. In the statistics literature, such difficulties would be viewed as censored data problems or sampling from truncated distributions. The term restriction of range tends to obscure different kinds of sample selection models. Consider three examples. (1) A company needs 50 employees to open a new division. It simply tests as many people as needed until 50 applicants with scores above some critical value are obtained. (2) A school wishes to recruit the best students. From those among the applicants, the school accepts only individuals with scores above some critical value, a value that may vary depending on the applicant pool. (3) Suppose a business interviews only candidates a consulting firm has selected for them, individuals with scores above some critical test value. The firm may have no knowledge about those applicants with scores too low to qualify for consideration.

The first two examples could be regarded as examples of censored data, because the total number of applications is known, and for all applicants' scores on the predictor test are known; scores on the criterion variable are known for the selected applicants and are unknown or censored for the unselected applicants. In addition, in the first example, the number of selected applicants is fixed, while in the second example it is random. In the third example, no knowledge of the larger applicant pool is available to the business and so this example may be regarded as sampling from a bivariate normal distribution with truncation from below on the predictor variable. If no use is made of the information available from unselected persons in the first two examples, then all three might be regarded as sampling from a truncated distribution. Of course, given the political and social realities of modern society, these examples are doubtlessly oversimplifications of reality.

Clearly, range restriction can be a very severe problem, and it may well be of much more importance than unreliability. Certainly, there are some dramatic real examples of the influence of range restriction (Linn, 1983). If selection is severe, then it may not be realistic to regard the mixture model of Equation (5.4.1) as having components $h(r)$.

Assume, as before, a bivariate normal model of prediction and criterion variables as having population correlation coefficient ρ when there is no selection. Let r_b denote the observed correlation coefficient computed on the selected subsample of those individuals with scores exceeding the critical value on the selection or predictor variable. Cohen (1955) shows that the maximum likelihood estimate of ρ is r_c, defined by

$$r_b/[1 - \hat{\kappa}(1 - r_b^2)]^{1/2} = r_c, \qquad 0 \leqq \hat{\kappa} < 1, \qquad (8.1.2)$$

where κ is equal to one minus the ratio of the variance of the restricted population to the variance of the unrestricted population, and $\hat{\kappa}$ is an estimate. Cohen gives different estimates of κ for different sampling schemes. One estimate is simply one minus the corresponding sample variance ratio, the familiar estimate.

The distribution of R_c is not known, but it might be expected to be approximately distributed as $h(r_c)$. Clearly, as κ approaches zero, $R_b = R_c$ in which case R_b will have the distribution of R. Thus, it might be hoped that R_c will in general have the distribution of R, at least approximately.

To provide some empirical evidence on the matter, a sampling scheme was employed that followed example (1) above. Values of X were sampled from a bivariate normal distribution with $\rho = .5$. If the sampled $x \geqq x_c$, a critical value, the corresponding value of y was sampled. If $x < x_c$, a new x was sampled. Sampling continued until 50 x and y pairs were obtained and then the correlation coefficient was computed, denoted in this sampling scheme by

FIGURE 8.2. The empirical distribution function for 3225 observed r_c for $n = 50$ with κ estimated from data. Population values were $\rho = \kappa = \frac{1}{2}$. $H(r)$ distribution function of R for $\rho = \frac{1}{2}$, $n = 50$.

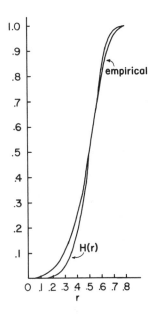

r_b. Thus, sampling might be regarded as restriction or censored sampling with censoring on y unless $x \geq x_c$. Because 50 (x, y) pairs were desired, the number of x values sampled was random (but at least 50), and x_c we set so that $\kappa = \frac{1}{2}$. Here, 3225 values of r_b were obtained in this way. Each r_b was corrected by Equation (8.1.2) and in two ways. First, κ was known, so in one case κ was fixed at $\kappa = \frac{1}{2}$ for all r_b. In the other case, κ was estimated for each r_b using the estimate noted above.

Figure 8.2 shows the empirical distribution function for r_c computed for either $\kappa = \frac{1}{2}$ or κ estimated. The differences in empirical probabilities under these two procedures were negligible with differences showing in the third decimal place. Figure 8.2 also graphs $H(r; .5, 50)$ for comparison. The empirical function follows H well except in the lower tail, where the correspondence is somewhat disappointing; it should be possible, however, to adjust Equation (8.1.2) so that the distribution of R_c would follow H more closely. Incidentally, the graph of the exact distribution of R from David's (1938) tables follows H almost exactly.

To see just how different the distribution of R_b can be from R, under the assumption that R_c has density $h(r_c)$ and with κ fixed, the density of R_b obtained by transformation is

$$h_2(r_b) = d(2\pi\sigma^2)^{-1/2} \exp\{-[z(r_b/v^{1/2}) - z(\rho)]^2/2\sigma^2\}, \qquad (8.1.3)$$

with $v = 1 - \kappa(1 - r_b^2)$, and $d = [v^{1/2}/(v - r_b^2)][1 - \kappa r_b^2/v]$.

Figure 8.3 illustrates distributions of $h_2(r_b)$ for $\kappa = \frac{15}{16}, \frac{3}{4}, \frac{7}{16}$, and zero. These values correspond to the ratios of the standard deviation of the selected to unselected populations of $\frac{1}{4}, \frac{1}{2}$, and $\frac{3}{4}$ and 1 for $\rho = .5$ and $n = 50$. The figure

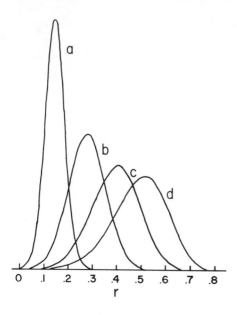

FIGURE 8.3. The distribution $h_2(r_b)$ with $\kappa = \frac{15}{16}, \frac{3}{4}, \frac{7}{16}$, and 0 for a, b, c, and d, respectively.

shows that range restriction can have a striking impact on the distribution of R_b, so that if range restriction is an important problem, to model R_b as if it had the distribution of R could lead to serious estimation errors.

8.1.4 CORRECTING FOR BOTH UNRELIABILITY AND RANGE RESTRICTION

As has been seen, range restriction can result in a random variable R_b with a distribution that is likely to be poorly modeled by $h(r_b)$. On the other hand, unreliability attenuates the population correlation coefficient so that ρ' is reduced to $\rho = \rho'\alpha\beta$, but otherwise leaves the distribution of R unchanged. This suggests that when correcting for both artifacts range restriction corrections should be applied first so that R_b is replaced by R_c, which may be modeled approximately by $h(r_c)$. Once range restriction corrections have been applied, reliability corrections may be considered.

Considering the distribution of the correlation coefficient under artifact corrections has assumed that the corrections κ and $\alpha\beta$ were known and fixed, and were not values of random variables. In practice, however, this assumption is never satisfied, and estimates or guesses of κ and $\alpha\beta$ are obtained. Viewing κ and $\alpha\beta$ as random variables complicates the picture and the above distribution results no longer hold. However, it seems doubtful that the additional model complexity necessary to deal with the distribution problem when these quantities are random would yield practical increments over the more elementary procedures outlined above. The simulation reported above provides support for this statement, although the problem should receive further study. The differences between viewing the correction factors as random or fixed would certainly be negligible if the sample size on which the estimates are based were reasonably large.

8.2 Identifiability of Mixtures

Identifiability can be a problem in mixture models. In general, however, the problem rarely arises when the components of the mixture are continuous densities (Everitt & Hand, 1981, p. 6). Furthermore, it is known that in the general univariate normal mixture case the parameters are identifiable (Titterington et al., 1985, p. 38).

The conditional mixture model of Equation (5.4.1) is essentially a normal mixture model; that is, the same procedures could have been proposed utilizing normal components from Fisher's Z transformation, where Z is defined as normal. If this route had been taken, the variance of Z would have been known, $(n - 3)^{-1}$, and thus the problem would be a normal mixture problem, but a simpler one because the component variance is known. Because R and Z in Equation (5.3.1) are in one-to-one correspondence, there is no reason to suspect any identifiability problem.

8.3 Failure of Model Assumptions

As Box (1976) observed, no model of real-world phenomena is a correct model, so the issue is not whether the model is correct or not. The issue is whether the model is importantly wrong. Consider each of these model assumptions.

8.3.1 R Is g DISTRIBUTED

It must be assumed that the predictor and criterion variables are bivariate normal in distribution, so that R has distribution $g(r)$, which may be approximated by $h(r)$.

Doubtlessly, bivariate normality rarely holds, at least not exactly in real data. For one reason, virtually all psychological measurements are positively distributed, and the normal distribution assigns positive probability to negative values. If normality holds for a population of job applicants, it cannot hold for the population defined by a nonrandom selection rule, such as rejecting all applicants with scores below some critical score.

The important issue is whether normality holds sufficiently well so that each R, whether corrected or not, is at least approximately distributed as $h(r; \rho, n)$. If not, then there will be model failure.

8.3.2 EACH (R, N) IS INDEPENDENT

At some point, independence is an almost inescapable assumption of model-based inference; if the assumption that each sample pair (R_i, N_i) is independent fails, then the joint density is not given by Equation (6.2.1).

Clearly, in many—perhaps most—real-world samples of validity coefficients, there is dependency among the coefficients. This is clearly evident in Schrader's (1977) law school data. It is difficult, however, to specify in general how lack of independence is likely to alter inference. The problem needs study.

From a practical standpoint, the analysis of subsets of independent validity coefficients, such as was employed for the law school example (Section 7.8), might be a sensible strategy. Suppose the solutions from independent samples were reasonably similar, and similar to a subsequent analysis of the entire data set. Such a finding might provide confidence that dependency, at least for the data set under consideration, did not seriously alter the picture.

8.3.3 MIXTURE PARAMETERS AT EACH n ARE IDENTICAL

At each sample size n, it is assumed that the number of components and the parameters of the conditional mixture distributions $h(r|n)$ are the same. Although a strong assumption, it is difficult to see why it should fail in most settings, but it is possible to envision how it might fail. Suppose for the most prestigious law schools class sizes were very small, and applicant test scores were very high. The correlational structure in such schools might well be different from the structure in less prestigious schools that might accept larger numbers of students with a more heterogeneous range of scores.

A very crude test of this speculative hypothesis might be to examine the magnitude of the validity coefficients as they relate to the sample size on which they are based. For the 150 data values of Example 6, Section 7.8, the correlation between the pairs (r_i, n_i) was .017, suggesting perhaps no obvious assumption failure.

Should the assumption be viewed as unreasonable, the separate analysis of homogeneous subsets of data might be a plausible alternative. The model could be altered to relax this assumption but it is unclear whether the additional model complexity would contribute usefully to inference.

8.4 Properties of the Maximum Likelihood Estimates

In general, the justification for ML estimates resides in their large sample properties. It is well known that ML estimates are, under a wide variety of conditions, consistent estimators, and thus the estimates converge to their population values as the sample size (s the number of correlation coefficients) increases (e.g., Norden, 1972; Everitt & Hand, 1981). However, these general results cannot be accepted uncritically in the present case because there are some special problems.

First, recall again that the conditional mixture density, Equation (5.4.1), with components $h(r)$, is not the exact distribution of R; $h(r)$ is only a close approximation. Consequently, if any convergence theorems are to be applicable, it must be assumed that R is precisely $h(r)$ in distribution, and of course, this assumption is wrong; hopefully, however, it is not importantly wrong.

Second, most theorems on consistency of the ML estimates assume that the sample values are iid. If the perspective taken is that pairs are of the form (r_i, n_i), $i = 1, \ldots, s$, are independent sample elements from the density $h(r, n)$ of Equation (5.4.3), the joint distribution of N and R, then they are iid with this density.

From this bivariate sampling perspective, given rather mild regularity conditions, theorems 3.1 and 3.2 of Redner and Walker (1984) provide the working justification for the ML estimates in the present case (and indeed for most mixture model estimates) for they guarantee consistency and asymptotic normality of the estimates $\hat{\rho}_j$ and $\hat{\lambda}_j$.

Viewing (R_i, N_i) as a random pair means that each n_i must be regarded as a value from some distribution of N. But, clearly, there must be, from this perspective, different distributions of N because there are vary large differences in the observed sample values, n, from different data sets as the examples in Chapter 7 illustrate. This fact must impact on the interpretations of the estimates provided and, consequently, may limit the generality of plausible inferences.

Ultimately, the wide applicability and generally well-behaved properties of ML estimates (Norden, 1972) and the degree to which the distribution $h(r)$ is close to $g(r)$, as argued in Section 5.3, must provide the justification for the use of the ML estimates proposed here.

8.5 Miscellaneous Comments

As noted in Section 1.3, a major goal of this research was to provide an alternative model formulation for the validity generalization problem. This goal has been achieved, and it appears fair to conclude that the proposed model offers a number of important advantages over the traditional validity generalization procedures. The inferences are model-based maximum likelihood; point estimates can be provided of the ρ parameters of interest, and approximate confidence intervals and tests may also be constructed.

From an estimation perspective, the estimation of t, the number of model components in the mixture, remains the major estimation problem, but in most data, both real and artificial that have been explored, the specification of t has seemed clear. The fact that the sample size, n_i, is known for each study is a critical advantage, the importance of which cannot be overemphasized, because the component variance is then essentially known. Consequently, solution uncertainty is substantially reduced.

Although the model proposed here was motivated by the validity generalization problem in personnel psychology, the approach may be useful in meta-analytical techniques, particularily with the current stress on using the sample correlation coefficient as a measure of effect size (Rosenthal, 1984). The model would need to be reparameterized because the quantities of focus in meta-analysis are different from the quantities of focus here, but the general estimation procedure could likely remain intact.

There are, of course, other possible models that could be proposed. It would be possible to develop a thoroughgoing Bayesian model if one is attracted to the subjectivity of such an approach. A more attractive alternative would appear to be an empirical Bayes approach, which is not at all Bayes, but might be viewed as an alternative to the maximum likelihood approach proposed here. A model in the spirit of Rubin (1980) might be the place to begin.

Appendix

-CONTROL-

```
[0]   CONTROL;IT
[1]   ⍝DEFINE VECTORS RV,NV,RHOV, & PV OF R VALUES, SAMPLE SIZES
      , STARTING
[2]   ⍝RHO VALUES AND LAMBDA WEIGHTS; RHOV & PV HAVE LENGTH T≥2.
      DEFINE NT
[3]   ⍝ NUMBER OF ITERATIONS & C SOLUTION TOLERANCE, E.G., NT←12
      , C←.001.
[4]   ⍝TYPE 'CONTROL' TO IMPLEMENT SOLUTION.
[5]   T←⍴PV
[6]   MV←0.5×(⍟((1+RHOV)÷(1-RHOV)))
[7]   ZV←0.5×(⍟((1+RV)÷(1-RV)))
[8]   SDV←(÷(NV-3))*0.5
[9]   MLLV←⍳0
[10]  IT←1
[11]  'NUMBER OF ITERATIONS = ',⍕NT
[12]  'STARTING LAMBDA VECTOR = ',⍕PV
[13]  'STARTING RHO VECTOR =',⍕RHOV
[14]  SIMPLE
[15]  MODLL
[16]  BR1:DMIXR
[17]  MODLL
[18]  'THIS IS ITERATION ',⍕IT
[19]  'MODEL -2×LOGLIKELIHOOD = ',⍕M2MLL
[20]  'LAMBDA VECTOR = ',⍕PV
[21]  'RHO VECTOR = ',⍕RHOV
[22]  'STOP WHEN ITERATIONS = ',⍕NT
[23]  'OR WHEN DIFFERENCE BETWEEN LIKELIHOODS ≤ ',⍕C
[24]  IT←IT+1
[25]  →((⍴MLLV)=1)/BR1
[26]  →((IT≤NT)∧((|(MLLV[⍴MLLV]-MLLV[(⍴MLLV)-1]))≥C))/BR1
[27]  '@@@@@@@ SUMMARY @@@@@@@'
[28]  'MLE LAMBDA VECTOR = ',⍕PV
[29]  'MLE OF RHO VECTOR = ',⍕RHOV
[30]  'MLE OF RHO UNDER 1 COMPONENT MODEL = ',⍕((*(2×MU))-1)÷((*
      (2×MU))+1)
      ∇
```

-DENR-

```
[0]   DENR
[1]   HR←(÷(((○2)*0.5)×SDV[I]))×(*(-(((ZV[I]-MV[J])*2)÷(2×(SDV[I]
      *2)))))×(÷(1-(RV[I]*2)))
      ∇
```

-DMIXR-

```
[0]    DMIXR;PJRV;J;I;COMPV
[1]    ⍝MAIN ROUTINE
[2]    J←1
[3]    PJRV←⍳0
[4]    I←1
[5]    BR3:COMPV←⍳0
[6]    BR1:DENR
[7]    COMPV←COMPV,(PV[J]×HR)
[8]    J←J+1
[9]    →(J≤T)/BR1
[10]   J←1
[11]   BR2:PJRV←PJRV,(COMPV[J]÷(+/COMPV))
[12]   J←J+1
[13]   →(J≤T)/BR2
[14]   ⍝FORMS P(1|R), P(2|R), FOR ONE VALUE OF R, J = 1, 2, . . .
       ,T
[15]   J←1
[16]   I←I+1
[17]   →(I≤ρRV)/BR3
[18]   PV←(+/(⍉(((ρRV),T)ρPJRV)))÷(ρRV)
[19]   MV←((ZV×(NV-3))+.×(((ρRV),T)ρPJRV))÷((NV-3)+.×(((ρRV),T)ρP
       JRV))
[20]   RHOV←((*(2×MV))-1)÷((*(2×MV))+1)
       ∇
```

-SIMPLE-

```
[0]    SIMPLE;M2LL
[1]    MU←(+/(ZV×(NV-3)))÷(+/(NV-3))
[2]    M2LL←((ρRV)×(⍟(O2)))+((+/(⍟SDV))×2)+(+/(((ZV-MU)*2)×(NV-3))
       )
[3]    M2LL←M2LL+(2×(+/(⍟(1-(RV*2)))))
[4]    '-2LOGLIKELIHOOD UNDER 1 COMPONENT MODEL = ',⍕M2LL
       ∇
```

-MODLL-

```
[0]    MODLL;I;J;THRV;LLV
[1]    ⍝CONSTRUCTS MIXTURE MODEL LOGLIKELIHOOD
[2]    LLV←⍳0
[3]    I←1
[4]    BR1:J←1
[5]    THRV←⍳0
[6]    BR2:DENR
[7]    THRV←THRV,PV[J]×HR
[8]    J←J+1
[9]    →(J≤T)/BR2
[10]   LLV←LLV,(⍟(+/THRV))
[11]   I←I+1
[12]   →(I≤ρRV)/BR1
[13]   M2MLL←¯2×(+/LLV)
[14]   MLLV←MLLV,M2MLL
       ∇
```

Notes

Chapter 1

1.1. "Validity generalization is a demonstration that a selection procedure or kind of selection procedure permits valid inferences about job behavior or job performance across given jobs or groups of jobs in different settings" (Society for Industrial and Organizational Psychology, Inc., 1987, p. 26).

Although the notion of validity generalization should be distinguished from the class of procedures advocated by Schmidt and Hunter (e.g., 1977), in fact their procedures often mean, in the literature, validity generalization.

1.2. For purposes here, a satisfactory model is a well-defined, notationally consistent structural equation with the important model properties explicitly stated. For example, random variables need to be distinguished from parameters and the values random variables take on. Bounds on the random variables need to made explicit, and the values of hypothesized moments need to be stated. In addition, the model specification needs to be distinguished from the problem of parameter estimation.

Chapter 2

2.1. Let X and Y be random variables with proper density functions with $a < X < b'$ and $a < Y < b$, $b' < b$. Let H_x and H_y be the cumulative lower tails. Then, $H_x(b') = 1$ and $H_y(b') < 1$, hence $H_x \neq H_y$.

2.2. Define a rectangular distribution: X and Y have a distribution on a rectangle if and only if their joint density $f(x, y) > 0$ on the set $\{(x, y): a < x < b, c < y < d\} = \mathcal{R}$, and $f(x, y) = 0$ elsewhere.

Theorem 2.1. *Let X and Y be bounded random variables on a region $\mathcal{C}$ of the plane with $f(x, y) > 0$ on this region and zero otherwise. Let the marginal densities $g(x)$ and $h(y)$ be positive on $a < x < b$ and $c < y < d$ and zero otherwise.*

1. *If X and Y are independent, they are distributed on a rectangle. An equivalent statement is 2.*

2. *If X and Y are not distributed on a rectangle, they are not independent; that is, they are dependent.*

PROOF. Since 1. and 2. are equivalent statements, proving one will prove the other; (1) will be proven. By hypothesis, $f(x, y) > 0$ if and only if $(x, y) \in \mathscr{C}$. By independence, $f(x, y) = g(x)h(y)$, and $g(x)h(y) > 0$ if and only if $(x, y) \in \mathscr{R} = \{(x, y): a < x < b, c < y < d\}$. To show $\mathscr{C} \equiv \mathscr{R}$, suppose $(x', y') \in \mathscr{C}$, then $f(x', y') = g(x')h(y') > 0$ if and only if $(x', y') \in \mathscr{R}$. Similarly, if $(x', y') \in \mathscr{R}$, then $g(x')h(y') = f(x', y') > 0$ if and only if $(x', y') \in \mathscr{C}$. Therefore, $\mathscr{C} \equiv \mathscr{R}$; thus, X and Y are distributed on a rectangle. Since 1. has been proven, equivalently 2. has been proven. $\Box$

2.3. The expression g distributed will henceforth refer to the distribution of R when the variables involved are bivariate normally distributed. This distribution is given by Patel and Read (1982, p. 311).

2.4.

Theorem 2.2. *Assume P and E in Equation* (2.3.1) *have positive variances and that Equation* (2.3.2) *holds, that is,* $\mathscr{E}(E \mid \rho) \equiv \mathscr{E}(E_\rho) = 0$ *for all ρ. Then the correlation between P and E is zero, i.e.,* $\mathrm{corr}(P, E) = 0$.

PROOF. Because $\mathscr{E}(E \mid \rho) = 0$ by hypothesis, $\mathscr{E}\mathscr{E}(E \mid \rho) = \mathscr{E}(E) = 0$. Consider the joint distribution of P and E and specifically the regression function of E on P. Since $\mathscr{E}(E \mid \rho) = 0$, the conditional mean of E is linear in ρ (cf. Hogg & Craig, 1978, p. 75). Therefore, $0 = \mathscr{E}(E \mid \rho) = \mathrm{corr}(P, E)[\mathscr{V}(E)/\mathscr{V}(P)]^{1/2}(\rho - \mathscr{E}(P))$, which is zero for all ρ if and only if $\mathrm{corr}(P, E) = 0$. $\Box$

Chapter 3

3.1. Let the random variable $X \in [a, b]$, then $0 \leq \mathscr{V}(X) \leq (a - b)^2/4$. To show this result translate X to $C \in [-d, d]$ and $b - a = 2d$. $\mathscr{V}(X) = \mathscr{V}(C)$ is maximized with $\mathscr{E}(C^2)$ maximum at $pd^2 + (1 - p)(-d)^2 = d^2$, independent of p. $[\mathscr{E}(C)]^2$ is minimum when $\mathscr{E}(C) = 0$ with $p = \frac{1}{2}$. So $\mathscr{V}(C)$ is maximum at $\mathscr{E}(C^2) = d^2$. Now, translate back to $[a, b]$, verifying bounds on $\mathscr{V}(X)$.

Chapter 4

4.1.

Theorem 4.1. *There does not exist a consistent estimator of a nonidentifiable parameter.* (*That is, if* $U_1(x, \mu_1) = U_2(x, \mu_2)$ *are distributions of X with parameters* $\mu_1 = \mu_2$, *then if there is a consistent estimator of* μ_1 *and* μ_2, *it must be that* $\mu_1 = \mu_2$. *If* $\mu_1 \neq \mu_2$, *they are values of the nonidentifiable parameter, and* U_1 *and* U_2 *are observationally equivalent distributions.*)

PROOF. Consider a random variable X with distribution $U_1(x, \mu_1) = U_2(x, \mu_2)$ with parameters μ_1 and μ_2. Let X_i, $i = 1$ to k, be a sample from U_1.

Suppose $T(X_1, X_2, \ldots, X_k) = T_k$ is a consistent estimator of μ_1 and μ_2, $\mu_1 \neq \mu_2$, with distribution function $F_k(t)$, which depends on k. For T_k to be a consistent estimator of μ_1 means T_k converges weakly to μ_1, so by definition of weak convergence for all $\varepsilon > 0$, $\lim_{k \to \infty} P(|T_k - \mu_1| < \varepsilon) = 1$. But, if the X_i are a sample from U_1, they are a sample from U_2, since $U_1 = U_2$. Thus, it must also be that for all $\varepsilon > 0$, $\lim_{k \to \infty} P(|T_k - \mu_2| < \varepsilon) = 1$. Then, following Hogg and Craig (1978, p. 186), it follows that

$$\lim_{k \to \infty} F_k(t) = 0, \qquad t < \mu_1$$

$$= 1, \qquad t > \mu_1 \qquad \text{(i)}$$

$$= 0, \qquad t < \mu_2 \qquad \text{(ii)}$$

$$= 1, \qquad t > \mu_2$$

and these equations cannot all hold for $\mu_1 \neq \mu_2$. For example, if $\mu_1 < \mu_2$, there exists a t_0 such that $\mu_1 < t_0 < \mu_2$ and by (i) $\lim_{k \to \infty} F_k(t_0) = 1$, and by (ii) $\lim_{k \to \infty} F_k(t_0) = 0$. Thus, the supposition is false; there cannot be a consistent estimator of an unidentified parameter. $\square$

Note that if $U_1 = U_2$ implies $\mu_1 = \mu_2$ then $\mu_1 = \mu_2$ is identifiable; also no restriction is placed on how T_k is defined.

Chapter 7

7.1. The sequence of decision rules was: (a) select the most recent year available; (b) select two successive combined years if data for a single year is unavailable; (c) select day school if it is reported separately from night school; and (d) select full-time if reported separately from part-time data. In most cases, data for 1973 or 1974, typically combined, were selected.

References

Aitkin, M.A. (1964). Correlation in a singly truncated bivariate normal distribution. *Psychometrika, 29,* 263–270.

Aitkin, M., & Wilson, G.T. (1980). Mixture models, outliers, and the EM algorithm. *Technometrics, 22,* 325–331.

Anastasi, A. (1986). Evolving concepts of test validation. In M.R. Rosenzweig & L.W. Porter (Eds.), *Annual review of psychology: Vol. 37* (pp. 1–15). Palo Alto, CA: Annual Reviews.

Arnold, S. (1981). *The theory of linear models and multivariate analysis.* New York: Wiley.

Baker, D.D., & Terpstra, D.E. (1982). Employee selection: Must every job test be validated? *Personnel Journal, 61,* 602–605.

Basu, A.D. (1983). Identifiability. In S. Kotz & N.L. Johnson (Eds.), *Encyclopedia of statistical sciences: Vol. 4* (pp. 2–6). New York: Wiley.

Berger, J.O. (1985). *Statistical decision theory and Bayesian analysis.* New York: Springer-Verlag.

Bozdogan, H. (1987). Model selection and Akaike's information criterion (AIC): The general theory and its analytical extensions. *Psychometrika, 52,* 345–370.

Box, G.E.P. (1976). Science and statistics. *Journal of the American Statistical Association, 71,* 791–799.

Burke, M.J. (1984). Validity generalization: A review and critique of the correlation model. *Personnel Psychology, 37,* 93–115.

Burke, M.J., Raju, N.S., & Pearlman, K. (1986). An empirical comparison of the results of five validity generalization procedures. *Journal of Applied Psychology, 71,* 349–353.

Callender, J.C., & Osburn, H.G. (1980). Development and test of a new model for validity generalization. *Journal of Applied Psychology, 65,* 543–558.

Carruthers, J.B. (1950). Tabular summary showing relation between clerical test scores and occupational performance. *Occupations, 29,* 40–52.

Clarke, L.E. (1975). *Random variables.* New York: Longman.

Cohen, A.C. Jr. (1955). Restriction and selection in samples from bivariate normal distributions. *Journal of the American Statistical Association, 50,* 884–893.

Cramér, H. (1946). *Mathematical methods of statistics.* Princeton, NJ: Princeton University Press.

David, F.N. (1938). *Tables of the ordinates and probability integral of the distribution of the correlation coefficient in small samples.* Cambridge, England: The University Press.

Dunnette, M.D., Rosse, R.L., Houston, J.S., Hough, L.M., Toquam, J., Lammlein, S., King, K.W., Bosshardt, M.J., & Keyes, M. (1982). *Development of an industry-wide electric power plant operator selection system.* Washington, DC: Edison Electric Institute.

Efron, B., & Tibshirani, R. (1986). Bootstrap methods for standard errors, confidence intervals, and other measures of statistical accuracy. *Statistical Science, 1*, 54–77.

Everitt, B.S. (1984). *An introduction to latent variable models.* London: Chapman and Hall.

Everitt, B.S., & Hand, D.J. (1981). *Finite mixture distributions.* London: Chapman and Hall.

Gayen, A.K. (1951). The frequency distribution of the product-moment correlation coefficient in random samples of any size drawn from non-normal universes. *Biometrika, 38*, 219–247.

Ghiselli, E.E. (1966). *The validity of occupational aptitude tests.* New York: Wiley.

Ghosh, B.K. (1966). Asymptotic expansions for the moments of the distribution of correlation coefficient. *Biometrika, 53*, 258–262.

Helm, W.E., Gibson, W.A., & Brogden, H.E. (1957). *An empirical test of problems in personnel classification research.* Personnel Research Board Technical Research Note 84, October.

Hogg, R.V., & Craig, A.T. (1978). *Introduction to mathematical statistics* (4th ed.). New York: Macmillian.

Hunter, J.E., Schmidt, F.L., & Jackson, G.B. (1982). *Meta-analysis: Cumulating research findings across studies.* Beverly Hills: Sage.

James, L.R., Demaree, R.G., & Mulaik, S.A. (1986). A note on validity generalization procedures. *Journal of Applied Psychology, 71*, 440–450.

Johnson, N.L., & Kotz, S. (1970). *Continuous univariate distributions-2.* New York: Houghton Mifflin.

Kendall, M.G., & Stuart, A. (1979). *The advanced theory of statistics: Inference and relationship. Vol. 2* (4th ed.). New York: Macmillian.

Kraemer, H.C. (1975). On estimation and hypothesis testing problems for correlation coefficients. *Psychometrika, 40*, 473–485.

Landy, F. (1985). *Psychology of work behavior* (3rd ed.). Homewood IL: Dorsey.

Lehmann, E.L. (1983). *Theory of point estimation.* New York: Wiley.

Lent, R.H., Aurbach, H.A., & Levin, L.S. (1971). Predictors, criteria, and significant results. *Personnel Psychology, 24*, 519–533.

Linn, R.L. (1983). The Pearson selection formulas: Implications for studies of predictive bias and estimates of educational effects in selected samples. *Journal of Educational Measurement, 21*, 33–47.

Linn, R.L., Harnisch, D.L., & Dunbar, S.B. (1981). Validity generalization and situational specificity: An analysis of the prediction of first-year grades in law school. *Applied Psychological Measurement, 5*, 281–289.

Lord, F.M., & Novick, M.R. (1968). *Statistical theories of mental test scores.* Reading, MA: Addison-Wesley.

Norden, R.H. (1972). A survey of maximum likelihood estimation. *International Statistical Review, 40*, 329–354.

Novick, M.R., & Jackson, P.H. (1974). *Statistical methods for educational and psychological research.* New York: McGraw-Hill.

Ord, J.K. (1972). *Families of frequency distributions.* New York: Hafner.

Osburn, H.G., Callender, J.C., Greener, J.M., & Ashworth, S. (1983). Statistical power of tests of the situational specificity hypothesis in validity generalization studies: A cautionary note. *Journal of Applied Psychology, 68*, 115–122.

Patel, J.K., & Read, C.B. (1982). *Handbook of the normal distribution*. New York: Marcel Dekker.

Pearlman, K. (1982). *The Bayesian approach to validity generalization: A systematic examination of the robustness of procedures and conclusions*. Unpublished doctoral dissertation, George Washington University, Washington, DC.

Pearson, K. (1894). Contribution to the mathematical theory of evolution. *Philosophical Transactions of the Royal Society, A, 185*, 71–140.

Raju, N.S. & Burke, M.J. (1983). Two new procedures for studying validity generalization. *Journal of Applied Psychology, 68*, 382–395.

Redner, R.A., & Walker, H.F. (1984). Mixture densities, maximum likelihood and the EM algorithm. *SIAM Review, 26*, 195–239.

Rosenthal, R. (1984). *Meta-analytic procedures for social research*. Beverly Hills: Sage.

Rubin, D.B. (1980). Using empirical Bayes technique in the law school validity studies. *Journal of the American Statistical Association, 75*, 801–816.

Sackett, P.R., Schmitt, N., Tenopyr, M.L., Kehoe, J., & Zedeck, S. (1985). Commentary on forty questions about validity generalization and meta-analysis. *Personnel Psychology, 38*, 697–798.

Sakamoto, Y., Ishiguor, M., & Kitagawa, G. (1986). *Akaike information criterion statistics*. Boston: Reidel.

Schmidt, F.L., Gast-Rosenburg, I., & Hunter, J.E. (1980). Validity generalization results for computer programmers. *Journal of Applied Psychology, 65*, 643–661.

Schmidt, F.L., & Hunter, J.E. (1977). Development of a general solution to the problem of validity generalization. *Journal of Applied Psychology, 62*, 529–540.

Schmidt, F.L., & Hunter, J.E. (1978). Moderator research and the law of small numbers. *Personnel Psychology, 31*, 215–232.

Schmidt, F.L., Hunter, J.E., Pearlman, K., & Hirsch, H.R. (1985). Forty questions about validity generalization and meta-analysis. *Personnel Psychology, 38*, 697–798.

Schmidt, F.L., Hunter, J.E., Pearlman, K., & Shane, G.S. (1979). Further tests of the Schmidt–Hunter Bayesian validity generalization procedure. *Personnel Psychology, 32*, 257–281.

Schmidt, P. (1983). Identification problems. In S. Kotz & N.L. Johnson (Eds.), *Encyclopedia of statistical sciences: Vol. 4* (pp. 10–14). New York: Wiley.

Schrader, W.B. (1977). *Summary of law school validity studies* (LSAC-76-8). Newtown, PA: Law School Admission Council.

Serfling, R.J. (1980). *Approximation theorems of mathematical statistics*. New York: Wiley.

Sharf, J.C. (1987). Validity generalization: Round two. *The Industrial–Organizational Psychologist, 25*, 49–52.

Society for Industrial and Organizational Psychology, Inc. (1987). *Principles for the validation and use of personnel selection procedures* (3rd ed.). College Park, MD: Author.

Soper, H.E., Young, A.W., Cave, B.M., Lee, A., & Pearson, K. (1917). On the distribution of the correlation coefficient in small samples. *Biometrika, 11*, 328–413.

Thomas, H. (1982). IQ, interval scales, and normal distributions. *Psychological Bulletin, 91*, 198–202.

Thomas, H. (1983). Familial correlational analysis, sex differences, and the X-linked gene hypothesis. *Psychological Bulletin, 93*, 427–440.

Titterington, D.M., Smith, A.F.M., & Makov., U.E. (1985). *Statistical analysis of finite mixture distributions.* New York: Wiley.

Uniform guidelines on employee selection procedures (1978). *Federal Register, 43* (166), 11996–12009.

Wilks, S.S. (1938). The large-sample distribution of the likelihood ratio for testing composite hypotheses. *Annals of Mathematical Statistics, 9,* 60–62.

Zedeck, S., Cascio, W. (1984). Psychological issues in personnel decisions. In M.R. Rosenzweig & L.W. Porter (Eds.), *Annual Review of Psychology: Vol. 35* (pp. 461–518). Palo Alto, CA: Annual Reviews.

Author Index

Subject Index